彩图 2-2　傣族的竹楼
C. P. 2-2　Yunnan Dai nation bamboo house

彩图 2-3　延安石窑洞
C. P. 2-3　Cave dwelling built by stone

彩图 2-4　云南丽江玉龙山景观
C. P. 2-4　Yunnan Lijiang Yulong

彩图 2-5　苏州小桥流水人家
C. P. 2-5　Suzhou houses by stream through small bridge

彩图 2-10　法国巴黎高层建筑的镜面玻璃幕墙
C.P. 2-10　Mirror curtain wall of the high-rise building, Paris, France

彩图 2-6　北京大学校园一角——中国传统园林树例
C.P. 2-6　A corner of Beijing Univ. campus—an example of Chinese park's pavilian

彩图 2-9　我国某大城市一角(20 世纪 80 年代)
C.P. 2-9　A corner of a big city in China(the 1980s)

彩图 5-18 美国的大跨充气薄膜自然采光屋顶
C.P. 5-18 Daylighting by a big span film inflated roof, USA

彩图 5-23 日本的太阳能电池瓦
C.P. 5-23 Japan solar cell tie

彩图 5-3 日本的安装在屋顶的大面积太阳能热水器
C.P. 5-3 A big area solar heater on roof, Japan

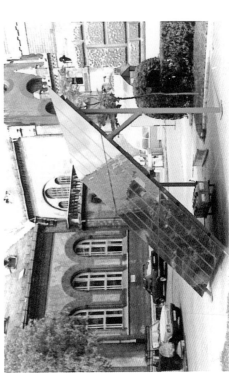

彩图 5-22 德国的太阳能电池板
C.P. 5-22 German solar cell pannel

彩图 5-25　法国某市政府地下会议室星点顶光天窗
C.P. 5-25　Daylighting with top star skylights in a city's government built underground, France

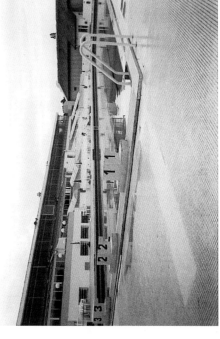

彩图 5-37　法国某地下热水露天游泳池
C.P. 5-37　An open swimming pool using the geothermal water, France

彩图 5-24　巴黎地下超级市场带形天窗采光
C.P. 5-24　Paris underground supermarket strip skylight

彩图 5-36　美国某公园露天太阳能游泳池
C.P. 5-36　An open solar swimming pool in a park, USA

彩图 7-1　美国明尼苏达大学地下建筑的地面绿化
C. P. 7-1　The ground greening of the Underground Space Centre, Minnesota Univ. USA

彩图 7-2　法国巴黎超级地下市场出入口之一
C. P. 7-2　An entrance of an underground supermarket in Paris, France

彩图 7-3　法国某地下空间上面为日光浴公园
C. P. 7-3　A sunbath park on an underground space, France

彩图 7-4(a)　法国某市政府地下建筑出入口
C. P. 7-4(a)　The entrance of a city government's underground building, France

彩图 7-4(b)　法国某市政府地下建筑屋顶绿化及下沉式温室
C. P. 7-4(b)　The roof greening and sunken greenhouse of a city government's underground building, France

彩图 7-5(a)　美国明尼阿波利斯市数据中心大型掩土太阳房
C. P. 7-5(a)　The big earth sheltered solar building of the control Data center, Minneapolis, USA

彩图 7-5(d)　前述太阳房向阳面太阳能集热器
C. P. 7-5(d)　The large area solar collector on the sunfacing side of the control Data center, Minneapolis, USA

彩图 7-6　日本一建筑师设想的日本深层地下城方案
C. P. 7-6　A deep underground building scheme designed by a Japan architect

彩图 7-7　联合国教科文组织总部(设在巴黎)地下办公楼
C. P. 7-7　The underground office buildings of UNESCO's Headquarters in Paris

彩图 7-8　巴黎市交通道旁掩土太阳房住宅覆土面绿化状况
C. P. 7-8　The greening view of the earth sheltered dwellings along a street in Paris

彩图 7-9　澳大利亚某掩土建筑卫星中继站
C. P. 7-9　The earth sheltered building of a satellite relay station, Australia

彩图 7-10(a)　日本川崎地下街一景
C. P. 7-10(a)　A view of the "Chigakai" in Kawasaki, Kanagawa, Japan

彩图 7-10(b) 图 7-10(a)的地下街锯齿天窗采光景观
C. P. 7-10(b) The skylight of C. P. 7-10(a)'s "Chigakai"

彩图 7-10(c) 日本某市地下街出入口
C. P. 7-10(c) The entrance of a city's "Chigakai"

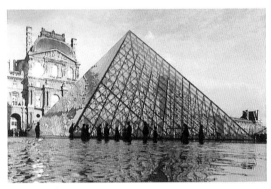

彩图 7-11 法国巴黎卢浮宫地下空间出入口锥形玻璃建筑
C. P. 7-11 The transparent pyramidic glass building is the entrance of the underground space of the plais du Louvre Paris, France

彩图 7-12(a) 法国某掩土建筑住宅
C. P. 7-12(a) An earth sheltered dwelling in France

彩图 7-13(a) 河南省洛阳郊区某下沉式窑洞村
C. P. 7-13(a) A sunken cave village in Luoyang suburb, Henan Province, China

彩图 7-13(b) 下沉式窑洞院落
C. P. 7-13(b) A sunken cave court

彩图 7-14(c)　延安大学石窑洞群
C. P. 7-14(c)　The stone cave group of Yanan University

彩图 7-17　饱含地方文脉的窑洞门窗(延安)
C. P. 7-17　cave's door & window fully contain local context

彩图 7-18　小窑洞——天然冷藏库
C. P. 7-18　The small cave-natural cold storage

彩图 7-31(a)　西安钟鼓楼广场地下商场屋顶绿化
C. P. 7-31(a)　The roof greening of Xi'an Bell-Drum Tower Square's underground shopping center

彩图 7-31(b)　西安钟鼓楼广场地下商场出入口
C. P. 7-31(b)　The entrance of xian Bell-Drum Tower square's underground shopping center

彩图 8-2(a)　佛山市绿化街景之一(佛山建委提供)
C. P. 8-2(a)　Foshan city street greening(1)
(Source：Foshan construction committee)

彩图 8-2(b) 佛山市绿化街景之二(佛山建委提供)

C. P. 8-2(b) Foshan city street greening (2) (Source: Foshan construction committee)

彩图 8-2(c) 深圳市绿化街景之一(孔令茜提供)

C. P. 8-2(c) Shenzhen city street greening (Source: Madam Kong Lingxi)

彩图 8-2(d) 深圳市住宅庭院绿化(于萍提供)

C. P. 8-2(d) Shenzhen city dwelling yard greening (Source: Madam Yuping)

彩图 8-2(e) 西安市新城广场绿化及市民聚此活动情况

C. P. 8-2(e) The greening and people action at the Xincheng Square of Xi'an city

彩图 8-2(f) 西安新城广场大面积草坪与花卉一景

C. P. 8-2(f) A view of the large grassland and flowers at the Xincheng Square of Xi'an city

彩图 8-3(a) 西安护城河水—绿结合视景之一
C. P. 8-3(a) Xi'an moat's water-green view (1)

彩图 8-3(b) 西安护城河水—绿结合视景之二
C. P. 8-3(b) Xi'an moat's water-green view (2)

彩图 8-4(a) 凡尔赛宫主建筑及广场水—绿结合视景
C. P. 8-4(a) The combinative landscape in the front of the main building and the entrance square of France Versailles Palace

彩图 8-4(*b*)、8-4(*c*)　凡尔赛宫水—绿—铜雕、石雕相结合一景
C. P. 8-4(*b*)、(*c*)　A combinative view of water-green-statues of bronze or stone in France Varsailles Palace

彩图 8-4(*d*)　凡尔赛宫水—绿结合一景
C. P. 8-4(*d*)　A water-green view of France Versailles Palace

彩图 8-4(*e*)　凡尔赛宫水—绿结合一景
C. P. 8-4(*e*)　Another water-green view of the versailles Palace

彩图 8-5　巴黎埃菲尔塔广场水—绿结合景观
C. P. 8-5　A water-green view of Eiffel Tower Square Paris

彩图 8-6　荷兰 HETLOO 公园水—绿—雕结合景观
C. P. 8-6　A water-green-statues view in HETLOO Park，Holland

彩图 8-7 新疆石河子市街道绿化
C. P. 8-7 The street landscape of Shihezi city, Xinjiang region

彩图 8-9 西安人民剧院西墙面绿化
C. P. 8-9 The greening of the west-facing wall of the Xi'an People Theatre

彩图 8-13 上海嘉定镇某宾馆屋顶绿化
C. P. 8-13 A hotel roof greening of Jiading Town of Shanghai city

彩图 8-14 西安建筑科技大学掩土太阳房屋顶及周围绿化
C. P. 8-14 Roof and surroundings' greening of a earth sheltered building in the campus of Xi'an Univ. of Arch. and Tech.

彩图 8-17 西安建筑科技大学散水种植绿化一例
C. P. 8-17 An example of apron planting, Xi'an Univ. of Arch. & Tech.

彩图 8-18　嘉定宾馆散水绿化（采用自由落水）

C. P. 8-18　The apron planting of Jiading Hotel (free-fall drainage)

彩图 8-24(a)　西安某阳台种植花卉一例（采用推拉窗）

C. P. 8-24(a)　Xi'an city a balcony planting (Sliding window)

彩图 8-24(b)　阳台向阳面牵牛花夏季遮阳视景

C. P. 8-24(b)　Morning glory on sun-facing balcony window as summer shading

彩图 8-25　深圳滨海宾馆窗台花盘绿化（刚种不久）

C. P. 8-25　Windowsill planting of a seaside hotel, Shenzhen city (It was planted a few days before)

彩图 8-28 西方古典建筑窗台花卉景例
C. P. 8-28 A windowsill flowers of a western classical building

彩图 8-31 住宅庭院与墙面绿化一例
C. P. 8-31 A greening example of a dwelling yard and wall

彩图 8-32 北京大学一教学楼墙面及屋顶绿化
C. P. 8-32 The greening of roof and wall of a teaching building in Beijing Univ.

彩图 8-33 匈牙利布达佩斯某使馆墙面绿化
C. P. 8-33 Wall greening of an embassy, Budapest

彩图 8-34 西安街头花卉制作的九龙壁
C. P. 8-34 A Nine dragon wall made of flowers of a street in Xi'an city

彩图 8-35　北京某建筑花卉围栏景色

C. P. 8-35　An enclosure made of flowers in Beijing city

彩图 8-36　深圳某花卉围栏景色

C. P. 8-36　An enclosure made of flowers in Shenzhen city

彩图 8-37　广州白天鹅宾馆室内绿化景色

C. P. 8-37　Indoor greening of White Swan Hotel in Guangzhou city

彩图 8-38　上海嘉定镇街道杉树夏季防西晒

C. P. 8-38　With firs for summer west shading of a street, Jiading town of Shanghai city

彩图 8-39　西安街道钻天杨夏季防西晒

C. P. 8-39　Summer west shading with tall poplars of a street in Xi'an city

彩图 8-40　清华大学阶梯教室东侧季青绿化防夏东晒

C. P. 8-40　Seasonal greening for summer east shading of a lecture theatre east side，Qinghua Univ.

彩图 8-41　巴黎圣母院主立面

C. P. 8-41　Facade of Notre Dame de Paris

彩图 8-42　好洁美的一座教堂

C. P. 8-42　What a pure white church!

高校建筑学 城市规划 景观专业系列教材

生态可持续建筑
ECO-SUSTAINABLE ARCHITECTURE(ESA)

（第二版）

（汉—英双语 Billingualism of Chinese-English）

西安建筑科技大学　　　夏云　主　编
北京工业大学　　　　　夏葵
西安交通大学　　　　　陈洋　副主编
万达文化旅游规划研究院　王进　英语总校对

中国建筑工业出版社

图书在版编目(CIP)数据

生态可持续建筑/夏云主编. —2版. —北京：中国建筑工业出版社，2012.1（2021.11重印）
（高校建筑学　城市规划　景观专业系列教材）
ISBN 978-7-112-13996-5

Ⅰ.①生… Ⅱ.①夏… Ⅲ.①生态建筑-可持续性发展-研究-英、汉 Ⅳ.①TU18

中国版本图书馆CIP数据核字(2012)第013147号

高校建筑学　城市规划　景观专业系列教材

生态可持续建筑
ECO-SUSTAINABLE ARCHITECTURE(ESA)
（第二版）
（汉—英双语 Billingualism of Chinese-English）

西安建筑科技大学	夏云	主　编
北京工业大学	夏葵	副主编
西安交通大学	陈洋	
万达文化旅游规划研究院	王进	英语总校对

*

中国建筑工业出版社出版、发行（北京西郊百万庄）
各地新华书店、建筑书店经销
北京天成排版公司制版
北京建筑工业印刷厂印刷

*

开本：787×1092毫米　1/16　印张：14$\frac{1}{2}$　插页：8　字数：379千字
2013年9月第二版　2021年11月第十五次印刷
定价：**38.00元**
ISBN 978-7-112-13996-5
(22054)

版权所有　翻印必究
如有印装质量问题，可寄本社退换
（邮政编码 100037）

本社网址：http://www.cabp.com.cn
网上书店：http://www.china-building.com.cn

本书是根据10年前的原书中文版本改写的，两大改进：一是采用汉—英双语；二是与时俱进进行了新陈代谢的改写，分10章：生态可持续发展的障碍；温故知今与创新；21世纪建筑何往；何谓生态可持续建筑；太阳能与沼气在建筑中的应用；增效资源；掩土建筑；建筑绿化；优化组合；生态可持续建筑的经济效益，对生态可持续建筑的理念、设计原理、方法、构造措施、材料选择进行了图文并茂的论证(述)。

本书适用于建筑学、城乡规划、景观园林、环境艺术、环境工程、土木工程等专业研究生及本科生的教材及相关设计、科研、管理及政策制定领域人员参考。

This book was rewritten based on the original one written in Chinese ten years ago. It has two major improvements: one is it is now bilingualism in Chinese and English; the other is that we updated many parts to reflect the latest information. The book is divided into ten chapters: Barriers of Eco-sustainable Development; Reviewing the Past to Understand Today and Creat the New; Where does Architecture Go in the 21st Century; What dose ESA mean; Multi-use of Solar Energy and Methane in Buildings; Increasing Resources Efficiency (IRE); Earth Sheltered Buildings; Greening of Buildings; Optimal Compositions; Economic Benefits of ESA. This book has demonstrated and discussed the ESA's ideas, design principles, methods, construction measures and selection of materials by paying equal attention on both pictures and words.

This book is suitable to the postgraduates, undergraduates of the specialisties of architecture, urban-rural planning, landscape & garden, environmental art, environmental engineering, civil engineering, etc. as teaching book, and as reference for all concerned in design, research, management and policy-making fields.

* * *

责任编辑：王玉容
责任设计：董建平
责任校对：肖　剑　刘　钰

中国工程院院士用笺

关于《生态可持续建筑》教科书的推荐信

1. 本书科学地论述了"生态"及"生态可持续建筑"的含义，全书贯注了"天地人和"的大环境观，提出："物我相应，和谐共生与环境友好"是建筑生态可持续发展的必由之路。完全正确。

2. 书中以一简明数学式揭示了增效资源/D效益与建筑平面体形结构、构造、总面布局以及建筑绿化等之间的关系；详尽论述了太阳能、沼气、风能、水能(SMWW)在建筑中的多种应用并根据大量实验提出了相应的设计原理与构造措施。对建筑及相关领域的生态可持续发展有现实的和有效的价值。

3. 医学专家张宜仁教授进行的细菌试验、生理实验、热环境比较、居住者问卷调查，证实掩土建筑不仅有节约建筑运行过程中能耗资源，还有节约居住者理疗耗能的习效，是未利于人体生命的因素。这是一项对掩土建筑宜居因素的新发现！

4. 作者在长期授课中受学生欢迎的双语教学模式，将本书从英语版改为汉－英双语版并进行了必时俱进的内容更新，符合教改的要求，符合全球21世纪交流大升温的趋势。

5. 全书既体现了：便于学生阅读、启发思考、激励创造性思维，增强学生在实践中分析、解决问题的能力——教材效果的初心改版。

我感佩深地推荐本书为"十二五"教材规划中等级教科书之一。

中国工程院院士 清华大学教授 博士导师
李道增 2010年7月20日

熊明的推荐信

评《生态可持续建筑》

该书是具有很高学术与实践价值的研究成果。

1. 全面深入系统地论述该课题在理论研究和实践方面的必要性和可行性。
2. 回顾和分析传统乡土建筑和现代城市建筑在生态环境方面的优缺点。
3. 提出21世纪建筑何往的课题,以及发展生态建筑的出路。
4. 太阳能及沼气的应用及技术。
5. 增效资源与城市形式、建筑布置,建筑构造的关系及其数学表达。
6. 掩土建筑的特点、类型、缺点及中外实例。
7. 建筑绿化的作用与构造设计。
8. 上述各种途径与方法的优化组合。
9. 最终达到目标:零—零建筑的构成。

总之,全书内容全面、系统;论述精辟,深入;逻辑分析清晰;文字简明、易懂;目标与途径明确。发人深省又授之以方法。建议作为大学教材及相关工作参考书籍正式出版。

北京市建筑设计研究院
顾问总建筑师　熊明教授
2010年7月30日

夏 云 简 介
(作者之一)

夏云教授(1927.12～)，1954～2010年教学56年。1950～1954年在北大、清华学习，1954年毕业于清华大学建筑学专业，留校任教至1957年。自1957年开始在西安建筑科技大学任教，曾兼任建筑学院教学督导及建筑物理实验室主任。

教学，汉英双语讲授：生态可持续建筑；建筑科学基础；建筑构造；建筑物理；建筑防火；专业英语。

访问讲学：国内16次16所院校(含4次外国学生)，国外3次3校、所。

出书：节能节地建筑基础；节能节地建筑(英语)；生态可持续建筑及建筑科学基础(汉—英双语)；英译汉两册；掩土建筑；建筑科学基础。

辅助教材：幻灯教材5集；建筑声学试验(英译汉35万汉字)。

科研：自1980年，论文60余篇(率研究生在12次相关国际会议上宣读28篇)。

专利：与女儿夏葵发明《太阳能水土热惯性富氧自然空调构造》获国家发明专利。(专利号：ZL99115793.1)

荣誉：部级科技进步二等奖一项；与他人合作获联合国教科文组织与国际建协主办的《为社区筹划可持续发展的未来》国际设计竞赛专业组第3名；国务院突出贡献特殊津贴；1999年被美国传略所(abi)选入世界名人录(Who's Who in the World)第16版；2002年，英国国际传略中心(IBC)授予"终身成就"奖；2007年，世界艺术、科学和信息大会再次授予"终身成就"奖。

Xia Yun Profile
(One of the authors)

Professor Xia Yun, (1927.12～) has been teaching for 56 years from 1954 to 2010. He has studied Architecture in Beijing University and Tsinghua University during 1950 to 1954 and graduated in 1954 from Tsinghua University and worked there till 1957. Since then Prof. Xia has been teaching in Xi'an University of Arch&Tech. He had been a part time Educational Advisor of the Architecture School and the head of building physics laboratory.

Teaching, following subjects bilingually in Chinese and English: Eco-sustainable Architecture; Foundations of Architectural Science; Building Construction; Building Physics; Building Fire Protection; Professional English.

Visiting lectures: Nationally in 16 colleges and universities (incl 4 times to international students) and internationally giving lectures abroad to 3 universities.

Publications: Books: Foundations of Energy & Land Saving Buildings; Energy & Land Saving Buildings (English); Eco-sustainable Architecture and Foundations of Architectural Science (bilingual in Chinese-English); Translation of two English books to Chinese:

Earth Sheltered Buildings; Foundations of Architectural Science.

Supplementary references: 5 sets of teaching slides; Building Acoustic Experiments (translation, English to Chinese 350 thousand Chinese words).

Researches: Since 1980, more than 60 papers (led postgraduates to present 28 papers in 12 international conferences/congresses).

Patent: With daughter Xia Kui invented "An Oxygen-rich Natural Air Conditioning Construction by Solar Energy & the Thermal Inertia of Water & Soil" which was patented in China (Patent number ZL99115793.1)

Honors: Awarded 2nd prize of Advanced Science and Tech of Ministry Level; With others won the 3rd professional prize of UNESCO & UIA, International Design Competition on "Designing for the Sustainable Future of a Community"; Special Subsidy of the State Council of China for Outstanding Contribution; in 1999 listed in *Who's Who in the World* (sixteenth edition), by US ABI; in 2002, "Lifetime Achievement Award" by IBC; in 2007, once more "Lifetime Achievement Award" by The World Congress of Arts, Sciences And Communications.

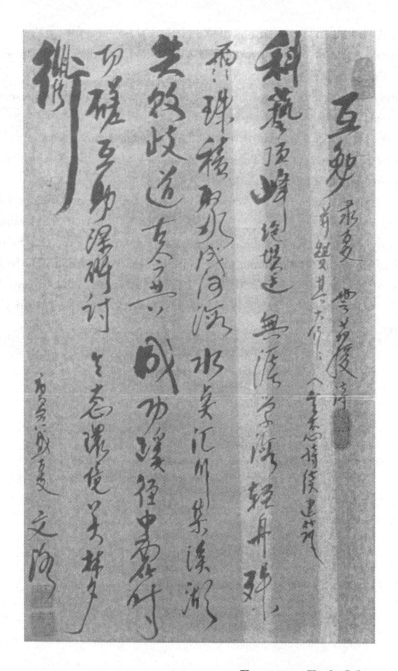

互勉

科艺顶峰绝坦途
学海无涯"奋"作舟
点滴积聚成深海
化纳千川乃汪洋
成就之道古今一
曲折成败相铺垫

本书作者们

Encourage Each Other

climb science - art top without any flat road
in boundless knowledge sea "self - rousing" is the boat
sea collecting waters drop by drop becomes deeper
Ocean welcomes and digests thousands of rivers to be vast
since ancient time the road to success is the same one,
that's based on twists, failures and successes

The authors

第二版前言

汉一英双语《生态可持续建筑》教科书的由来

当代，大学教育工作者与学生正关注两大潮流：一是所有发达国家和发展中国家都已经并正在研究、发展、实行其可持续发展战略；二是双语交流信息。许多国家已设立学汉语的孔子学院。联合国教科文组织（UNESCO）已设立孔学奖鼓励学汉语。全世界已有2500多所学校设有汉语课。我们当然更应学习有关外语。这两股潮流促使我们下决心应尽力使双语教材进入课堂，这就是我们要将原编汉语版《生态与可持续建筑》（2001.6，建工出版社）的教科书改编为汉一英双语《生态可持续建筑》新教材的由来。

我们是头一次编写这本双语教材，真诚希望得到批评指正以利改进。

在此除对第一版中提到的单位和个人再次表示感谢外，特对下列单位与个人在改编本书中给予的支持与帮助深表感谢：

感谢西安建筑科技大学建筑学院一贯支持；
感谢资深英语教师田一凡教授的热心帮助；
感谢剑桥大学海伦·茉莉根博士热心鼎力对重点章节的仔细校阅；
感谢文涛博士长期热心帮助收集资料；
感谢冯海燕女士大力承担稿本电脑排稿；
感谢医学专家张宜仁教授领导有关医学试验及校阅；
感谢责任编辑王玉容高工及其他相关编审者的热心帮助；
·本书作了多处新陈代谢的改写。

夏云

2011年05月04日，于西安

Second Edition Preface

The Birth of Chinese-English Bilingual Textbook ECO—SUSTAINABLE ARCHITECTURE (ESA)

Today, the university educators and students are facing two prominent trends: One is that both developed countries and developing countries have been studying, developing and implementing their sustainable development strategies. The other is the need of the bilingual exchange of information. Many countries have founded the Confucian Institutes to learn Chinese. The UNESCO has founded Confucian prize to encourage learning Chinese. Throughout the world more than 2500 schools have established Chinese course. On the other hand, we need to learn more relevant foreign languages. The two trends encouraged us to try our best to bring bilingual books into the class. That's why we want to rewrite the textbook Ecology and Sustainable Architecture (Written in Chinese, in June 2001, CBI Press) to a new teaching book Eco - Sustainable Architecture, in both Chinese and English.

It is our first endeavor in writing this bilingual teaching book. We whole-heartedly welcome any suggestion and comments so that mistakes can be corrected.

Here, besides thanks again to those whose names were mentioned in the original book written in Chinese we would specially express our gratitude to the following institutions and individuals for the supports and helps in rewriting the book:

College of Architecture, Xi'an University of Arch and Tech for consistent supports;

Senior English teacher Prof Tian Yifan for his kind help;

Cambridge University PHD Helen Mulligan for her unreserved help to carefully correct the important chapters;

Dr. Wen Tao for her long time heartly help to collect materials;

Madam Feng Haiyan diligently undertaking computerized publication of the manuscript;

Medical expert prof Zhang Yiren, for her leading relative medical tests and help in correction.

Editor-on-duty, senior engineer Wang Yurong and other editors concerned hearty help.

Many updates have been incorporated in this book.

Xia Yun
May 4tl, 2011, Xi'an

第一版前言

生态与可持续发展并不是新概念。万事万物与生俱来就固有地存在着这个问题，只是一直未被人们自觉地重视而已。近代人类的负效行为与天灾使人口膨胀、能源、土地、住房、环境等危机空前地危害到人类自身的生态与可持续发展，全球各行业才被惊醒，急求对策。建筑领域也不例外。从20世纪70年代起，对建筑的功能要求从传统的"庇护"发展到节能建筑，再到节能节地建筑，直到现在的生态与可持续建筑。认识到建筑及其环境必须具备有利于生态与可持续发展的物质与精神基础。为此，首先就得加紧研究，并在有关大专院校开出课程。

感谢国家自然科学基金会、西安建筑科技大学、陕西省建设厅、西安市科委、西安人防办、解放军总后西安建研所、日本太阳能学会、匈牙利太阳能学会等单位的支持与资助，使我们有条件进行研究，并取得较宽厚的编写本书的科研基础。已发表了《生态建筑与建筑的持续发展》、《自然空调掩土太阳房》等40余篇论文（其中28人次在12次国际会议上宣读28篇，并收入论文集），其公开成果有国外10余种杂志和专著引用或刊登。有的成果获部级科技进步二等奖，多篇获省、市优秀论文奖。夏云与他人合作参与联合国教科文组织与国际建协主办的"为可持续未来进行策划"的国际设计竞赛，获专业组第三名。

《生态与可持续建筑》课是1985年在西安建筑科技大学建筑系开设的《节能节地建筑》课的基础上，于1995年开出的，并首先被清华大学等10余所院系邀请作汇报讲学。使本书编写有了较坚实的教学实践基础。在该教材的编写过程中，张永录、姚远二位教授提供了科技史料考证；熊明大师惠赠新作；吴红、王朝晖、宋晔皓、戴彦军惠赠最新论文；佛山市建委、焦小浣、于萍、孔令茜、鲁渊理、加藤羲夫、让·鲁伯、海伦、格林兰、格兰尼提供了信息、资料。特别是本书医学顾问张宜仁教授不畏寒暑，领导本书的有关医学调查和试验工作。王玉容编审对书稿进行了非常认真仔细的审阅、改错，并热心提供正确编写教材的经验。

本书由周若祁教授积极推荐，全国建筑学专业指导委员会通过，建设部批准，最后由中国建筑工业出版社正式出版。因此，在该书即将问世之际，感谢建筑学专业指导委员会、建设部有关领导的热情关怀，感谢西安建筑科技大学及兄弟院校的大力支持，感谢上述同仁、朋友的大力协助。

在本书的使用过程中可能还会发现各种各样的不足及不妥之处，诚望批评指正。

夏 云
2000年10月12日
于西安

Preface

(First Edition)

Ecological and sustainable development is not a new concept. It has been existed in every matters' life cycle, It has not been consciously valued for a long time until recent years. Human negative behavior and natural disasters have made everyone waken up to the reality that the rapid population expansion, crises of energy, land, housing and environment are unprecedentedly damaging the ecological and sustainable development of ourselves. Professionals in all fields are working towards sustainable solutions. So is the architectural profession. Since 1970s, demands on buildings have expanded from being simple shelters to energy-saving, energy and land saving to the current ecologically sustainable. It is now a common understanding that buildings and their environment must form physical and social basis to facilitate the ecological and sustainable development. Consequently, we have to speed up the research in this area and set up relevant tertiary courses in the universities.

Our gratitudes go to Chinese Natural Science Research Funding, Xi'an University of Architecture and Technology, Shaanxi Bureau of Construction, Xi'an Science Committee, Xi'an Civil Air Defence Facility Office, Xi'an Building Research Institute of PLA Logistics Department, Japanese Solar Energy Institute, Hungarian Solar energy Institute for their generous support to fund our research. We have published over 40 papers such as "Ecological Architecture and Sustainable Building Development", "Natural Air Conditioned Earth-sheltered Buildings", etc. (among which 28 papers have been presented in 12 international conferences and compiled in the proceedings concerned), and a part of these works also published in over 10 profesional journals and books internationally. One paper has won 2nd Advanced Science and Technology prize of Ministry level. Many of them are awarded excellence in Provincial appraisal. Xia Yun, collaborated with others, has participated and won 3rd prize in professional category in an international design competition "Planning For A Sustainable Future", hosted by UNESCO and UIA.

The course of "Ecological and Sustainable Architecture (ESA)" began in 1995 in Xi'an University of Arch and Tech. It was based on the earlier course of "Energy Saving and Land Saving Buildings" in 1985 in the Dept. of Arch. of the same university. The course of ESA has been a series of lectures to Tsinghua University and 10 other universities. The above experiences form a solid base of this book. During the process of writing this book, Professor Zhang yonglu and yao yuan have provided solid proof in area of science history. Master Xiong Ming has kindly provided his new books. New research papers are provided by Wu Hong, Wang Zhaohui, Song Yehao, Dai Yanjun. Valuable information and document are supplied by Foshan City Building Committee, Jiao Xiaohuan, Yu Ping, Kong Lingqian, Lu Yuanli, 加藤義夫, Jean Paul Loube, Helen Mulligan, Jack Greenland, and G·Glaney. Special thanks to the medical research advisor of this book, Prof. Zhang Yiren, She has led relevant medical research and testing works regardless harsh weathers. Editor Wang Yurong has tirelessly checked and proof read the works and kindly provided invaluable knowledge on compiling such books.

This book is highly recommended by Prof. Zhou Rouqi. It is also examined by the National Architectural Advisory Committee, proved by the Ministry of Construction and finally published by the Construction Industry Press, At the time of its official debut, we wish to express our gratitude to the interest from all above parties, great support from Xi'an University of Arch. and Tech. and collaborating universities, and kind helps from above-mentioned colleagues and friends.

We would appreciate your advice and corrections should you find out any errors of omissions in this book.

Xia Yun
2000/10/12   Xi'an City

本书撰写人分工：

前言、第6章 ·· 夏云、施燕
第1、2章，第7章中7.3.4，7.3.5节 ···················· 夏葵
第3、5章 ·· 陈洋
第4、7章 ·· 靳亦冰
第8、9、10章 ·· 陈晓育
英语总校对 ·· 王进
全书由夏云统稿

Compilers' Division of Labour

Preface, Chapter 6 ·· Xia Yun, Shi Yan
Chapters 1, 2 and sections 7.3.4, 7.3.5 In Chapter 7 ········· Xia Kui
Chapters 3, 5 ··· Chen Yang
Chapters 4, 7 ··· Jin Yibing
Chapters 8, 9, 10 ·· Chen Xiaoyu
English Proofreading ············ (Senior architect, Australian Chinese) Wang Jin
Collection of all chapters ···································· Xia Yun

目 录
CONTENTS

1 生态可持续发展的障碍 ·· 1
1 BARRIERS OF ECO-SUSTAINABLE DEVELOPMENT ····················· 1
  1.1 人口激增 ··· 1
  1.1 Population Explosion(Population Boom) ··· 1
    1.1.1 膨胀的出生率 ··· 1
    1.1.1 Expanded birth rate ·· 1
    1.1.2 膨胀的需求 ·· 2
    1.1.2 Expanded needs ·· 2
    1.1.3 人类必须自控 ··· 2
    1.1.3 Humankind must be self-controlled ··· 2
    1.1.4 变人口负担为人才资源 ·· 2
    1.1.4 To make population from being a load to be a talent resource ········ 2
  1.2 自然灾害 ··· 3
  1.2 Natural Disasters ··· 3
    1.2.1 大自然灾害 ·· 3
    1.2.1 Serious natural disasters ·· 3
    1.2.2 其他自然灾害 ··· 5
    1.2.2 Other natural disasters ·· 5
  1.3 人类负效行为 ·· 5
  1.3 Human Negative Behaviours ··· 6
    1.3.1 人类的严重负效行为 ·· 5
    1.3.1 Man's serious negative behaviours ··· 6
    1.3.2 其他人为危害 ··· 10
    1.3.2 Other harms by man ·· 10

2 温故知今与创新——中国建筑回顾 ··· 11
2 LOOKING BACK TO THE PAST TO UNDERSTAND TODAY AND CREATE
   THE NEW——Reviewing the Chinese Architecture ······························· 11
  2.1 乡土建筑 ··· 11
  2.1 Vernacular Architecture ·· 11
    2.1.1 概况 ·· 11
    2.1.1 General ··· 11

  2.1.2 乡土建筑的优点 ……………………………………………………… 12
  2.1.2 Advantages of vernacular architecture ……………………………… 13
  2.1.3 乡土建筑的缺点 ……………………………………………………… 13
  2.1.3 Disadvantages of vernacular architecture …………………………… 13
 2.2 城市建筑 ………………………………………………………………… 14
 2.2 Urban Architecture ……………………………………………………… 14
  2.2.1 引言 …………………………………………………………………… 14
  2.2.1 Introduction …………………………………………………………… 14
  2.2.2 城市建筑优点 ………………………………………………………… 15
  2.2.2 Advantages of urban architecture …………………………………… 15
  2.2.3 城市建筑缺点 ………………………………………………………… 15
  2.2.3 Disadvantages of urban architecture ………………………………… 15
 2.3 新概念建筑 ……………………………………………………………… 16
 2.3 New Concept Architecture ……………………………………………… 17
  2.3.1 引言 …………………………………………………………………… 16
  2.3.1 Introduction …………………………………………………………… 17
  2.3.2 能源危机引发了节能建筑与太阳房的研究与推广 ………………… 17
  2.3.2 Energy crisis led to the research and implementation of energy saving buildings and solar buildings …………………………………………… 17
  2.3.3 绿色与土地危机引发了节能节地建筑 ……………………………… 20
  2.3.3 Green and land crises led to energy and land saving buildings ……… 20
  2.3.4 生态可持续危机引发了生态可持续建筑 …………………………… 22
  2.3.4 The crisis of eco-sustainability led to eco-sustainable architecture …… 22

3 21世纪建筑何往 ……………………………………………………………… 25
3 WHERE DOES ARCHITECTURE GO IN THE 21ST CENTURY ………… 25
 3.1 必然与偶然的反思 ……………………………………………………… 25
 3.1 Looking Back on Necessity and Contingency ………………………… 25
 3.2 建筑含义的延伸 ………………………………………………………… 26
 3.2 Extending the Meaning of Architecture ………………………………… 26
 3.3 21世纪建筑前景 ………………………………………………………… 27
 3.3 Outlook of Architecture in 21st Century ………………………………… 27
  3.3.1 上天——发展太空建筑 ……………………………………………… 27
  3.3.1 Go into sky-to develop space architecture …………………………… 27
  3.3.2 下海——发展海洋建筑 ……………………………………………… 31
  3.3.2 Go into the sea-to develop marine architecture ……………………… 32
  3.3.3 陆地建筑仍是主要方向 ……………………………………………… 32
  3.3.3 Land architecture-still the main direction …………………………… 33

4 何谓生态可持续建筑 ………………………………………………………… 34

| 4 | WHAT DOES ECO-SUSTAINABLE ARCHITECTURE MEAN | 34 |

4.1 可持续性并不是新概念 …… 34
4.1 Sustainability is not a New Concept …… 34
4.2 非可持续建筑 …… 35
4.2 Non-sustainable Architecture …… 35
4.3 可持续建筑含义 …… 36
4.3 The Meaning of Sustainable Architecture …… 37

5 太阳能与沼气在建筑中的应用 …… 45
5 THE APPLICATION OF SOLAR ENERGY AND METHANE IN BUILDINGS …… 45
　5.1 太阳 …… 45
　5.1 The Sun …… 45
　　5.1.1 太阳基本情况 …… 45
　　5.1.1 About the sun …… 45
　　5.1.2 应用太阳能的10个"最" …… 45
　　5.1.2 The ten "most" of using solar energy …… 46
　　5.1.3 中国太阳能资源 …… 46
　　5.1.3 The solar energy resources in China …… 46
　5.2 建筑中应用太阳能 …… 47
　5.2 Application of Solar Energy in Buildings …… 48
　　5.2.1 太阳能热水与炊事 …… 47
　　5.2.1 Solar hot water and cooking …… 48
　　5.2.2 中国古代太阳能集热器 …… 54
　　5.2.2 The ancient Chinese solar collectors …… 55
　　5.2.3 多功能构件 …… 55
　　5.2.3 Multi-function members …… 56
　　5.2.4 太阳能采光与日照 …… 57
　　5.2.4 Day lighting and sunshine …… 57
　　5.2.5 太阳能建筑 …… 67
　　5.2.5 Solar buildings …… 67
　5.3 沼气 …… 79
　5.3 Marsh Gas …… 80
　　5.3.1 沼气基本知识 …… 79
　　5.3.1 Basic knowledge of marsh gas …… 80
　　5.3.2 沼气池构造及工作原理 …… 81
　　5.3.2 Construction and operating principle of methane-generating pit …… 81

6 增效资源 …… 84
6 INCREASING RESOURCES EFFICIENCY (IRE) …… 84
　6.1 分析"F" …… 85

6.1　Analyzing "F" ·· 85
　6.1.1　从"F"看建筑体形与增效资源 ·· 85
　6.1.1　From "F" to Understand Building Shape and IRE ······································· 85
　6.1.2　太阳能建筑体形系数 ··· 88
　6.1.2　The Building Shape Coefficient(BSC) of solar buildings ··························· 88
　6.1.3　从"F"看建筑结构、构造与增效资源 ·· 88
　6.1.3　From "F" to understand building structure, construction and IRE ············· 89
　6.1.4　从"F"看建筑布置与增效资源 ·· 89
　6.1.4　From "F" to understand building layout and IRE ····································· 89
　6.1.5　从"F"看城市形式与增效资源 ·· 92
　6.1.5　From "F" to understand city's form and IRE ··· 93
6.2　分析"R" ·· 93
6.2　Analyzing "R" ·· 94
　6.2.1　实有热阻 R_p ·· 93
　6.2.1　Practical envelope thermal resistance R_p ··· 94
　6.2.2　节能热阻 R_{ES} ·· 97
　6.2.2　Energy saving thermal resistance R_{ES} ··· 97
　6.2.3　复合墙剖面设计 ·· 103
　6.2.3　Section design of composite wall ·· 104
　6.2.4　复合墙增热阻不增重或微增重 ·· 104
　6.2.4　To increase CW's thermal resistance without adding any weight or slightly increasing weight ··· 105
　6.2.5　旧墙体改造 ··· 105
　6.2.5　To remodel old walls ··· 105
6.3　分析"Z"与"t_o" ··· 107
6.3　Analyzing "Z" and "t_o" ··· 107
　6.3.1　科学安排"Z" ·· 107
　6.3.1　Scientifically managing "Z" ·· 107
　6.3.2　人能改变室外气温"t_o"吗? ·· 107
　6.3.2　Can the outdoor temperature "t_o" be changed manually? ······················· 108
6.4　分析"Q"与"$SRHG$" ··· 109
6.4　Analyzing "Q" and "$SRHG$" ··· 110
　6.4.1　反射隔热 ·· 110
　6.4.1　Heat insulation by reflection ·· 111
　6.4.2　通风 ··· 113
　6.4.2　Ventilation ··· 114
　6.4.3　阻存 ··· 115
　6.4.3　Resistance and storage(RS) ··· 115
　6.4.4　遮阳 ··· 116
　6.4.4　Sunshade ··· 117

6.4.5		绿化	118
6.4.5		Landscaping	119
6.4.6		自然空调	119
6.4.6		Natural air conditioning(NAC)	119
6.4.7		多功能构件	119
6.4.7		Multi-function members	119
6.4.8		多功能窗	119
6.4.8		Multi-function window	120
6.5		本章小结	126
6.5		Conclusion of the chapter	126

7　掩土建筑 ································ 130
7　EARTH SHELTERED BUILDINGS ·············· 130

7.1	引论	130
7.1	Introduction	130
7.2	国外掩土建筑实例	133
7.2	Examples of Earth Sheltered Buildings Internationally	133
7.3	中国窑洞	139
7.3	Chinese Caves(Yao dong)	139

7.3.1	中国窑洞的特点	139
7.3.1	The characteristics of Chinese caves	139
7.3.2	中国窑洞类型	141
7.3.2	Types of Chinese Caves	141
7.3.3	中国老窑洞缺点与综合治理	144
7.3.3	Old Chinese caves' shortcomings and comprehensive management	144
7.3.4	自然空调系统	147
7.3.4	Natural air conditioning system (NACS)	148
7.3.5	掩土建筑卫生状况	152
7.3.5	Hygiene condition in earth sheltered houses	152

7.4	中国城市地下空间	160
7.4	Underground Spaces in Chinese Cities	160

7.4.1	城市地下空间概况	160
7.4.1	General picture of underground spaces in cities	160
7.4.2	城市地下空间一范例	161
7.4.2	A good example of underground space in city	162
7.4.3	小结	162
7.4.3	Summary	162

8　建筑绿化 ································ 164
8　GREENING OF BUILDINGS ················ 164

8.1　光合作用的伟大贡献 ……………………………………………………… 164
8.1　The Great Contributions of Photosynthesis ……………………………… 164
8.2　建筑绿化的含义・效益・实例・实验 ……………………………………… 165
8.2　The Meaning・Benefits・Examples・Experiments of Greening of Buildings ……… 165
 8.2.1　含义・效益・实例 ………………………………………………… 165
 8.2.1　Meaning・benefits・examples …………………………………… 165
 8.2.2　阳台冬季温室效应实验 …………………………………………… 169
 8.2.2　Balcony greenhouse effect experiment in winter ……………… 169
 8.2.3　种植温室与日光间夏季比较 ……………………………………… 169
 8.2.3　Comparing planted greenhouse with sunspace in summer ……… 169
 8.2.4　小结 ………………………………………………………………… 172
 8.2.4　A brief summing-up ……………………………………………… 172
 8.2.5　种植温室与日光间冬季比较 ……………………………………… 173
 8.2.5　Comparing planted greenhouse with sunspace in winter ……… 173
8.3　建筑绿化构造设计 …………………………………………………………… 173
8.3　Construction Design of Greening of Buildings …………………………… 174
 8.3.1　屋面绿化构造 ……………………………………………………… 173
 8.3.1　Roof greening construction ……………………………………… 174
 8.3.2　散水绿化构造 ……………………………………………………… 177
 8.3.2　Apron greening construction …………………………………… 177
 8.3.3　窗台、阳台绿化 …………………………………………………… 183
 8.3.3　Windowsill and balcony greening ……………………………… 183
 8.3.4　墙面、柱面、挑檐面绿化 ………………………………………… 186
 8.3.4　Surface greenings of wall, column and overhang eaves ……… 186
 8.3.5　围墙绿化 …………………………………………………………… 186
 8.3.5　Greening of enclosure …………………………………………… 186
 8.3.6　室内绿化 …………………………………………………………… 187
 8.3.6　Indoor greening ………………………………………………… 187
 8.3.7　遮阳绿化 …………………………………………………………… 187
 8.3.7　Greening of sunshade …………………………………………… 187
 8.3.8　不是一切建筑都绿化 ……………………………………………… 188
 8.3.8　Not all buildings must be greened ……………………………… 188

9　优化组合 ……………………………………………………………………… 189
9　OPTIMAL COMPOSITIONS …………………………………………………… 189
9.1　能量自足房 …………………………………………………………………… 189
9.1　Energy Self-sufficient House ……………………………………………… 189
9.2　透明保热墙 …………………………………………………………………… 190
9.2　Transparent Insulated Wall (TIW) ………………………………………… 190
9.3　透明保热层多种用途 ………………………………………………………… 191

9.3	Multi-use of TIM	192
9.4	自控光热屋顶	192
9.4	Weather Panel Roof	192
9.5	地面掩土太阳房	194
9.5	Aboveground Earth Sheltered Solar Building	194
9.6	下沉式太阳能教室	195
9.6	Sunken Solar Classroom	196
9.7	下沉式温室自然空调太阳房	196
9.7	Sunken Greenhouse Solar Building with NAC	197
9.8	零—零建筑（双零建筑）	199
9.8	Zero—zero Building (Double Zero Building)	199
10	**生态可持续建筑的经济效益**	**203**
10	**ECONOMIC BENEFITS OF ESA**	**203**
10.1	回收与净得益	203
10.1	Benefit and Net Gain	203

主要参考文献 205
MAIN CONSULT DOCUMENTS 205

1 生态可持续发展的障碍
1 BARRIERS OF ECO-SUSTAINABLE DEVELOPMENT

什么叫"生态"?"生态"就是事物起源、生存、发展的条件与环境。人类也是宇宙事物之一,其生态可持续发展的条件与环境已经并正遭受着空前的危机。

人类自身的猛增,其负效行为以及无情的天灾乃是人类生态可持续发展的根本障碍。

What does "ecology (eco-)" mean? Ecology means the conditions and environments of things' origin, existence and development. Humankind is one of the universal beings, whose eco-sustainable developing conditions and environments have been suffering in unprecedented crisis.

Human population explosion, their negative behaviours and the ruthless natural disasters are the essential barriers obstructing man's eco-sustainable development.

1.1 人口激增

1.1.1 膨胀的出生率

世界人口已超 70 亿,每秒出生 4 人多,每 4 天 150 万,每年净增 8000 多万。

中国人口已超 14 亿,每 4 天出生 20 多万,每 3 年多,出生人口总量就相当于法国现有人口总量 6000 多万!

到 2050 年,世界人口如果控制得好,可能达到 100 亿左右,否则可能达 120 亿甚至更多!

1.1 Population Explosion(Population Boom)

1.1.1 Expanded birth rate

The world population is now over 7.0 billion. Every second there are more than 4 new born, each 4 days 1.5 million and every year the net increasing population is over 80 million.

Chinese population is over 1.4 billion. Each 4 days more than 0.2 million babies were born. In just over three years, the Chinese total new born population is equal to the existing total population of over 60 million in France!

To 2050, the world population, if could be controlled well, will reach 10.0 billion, if

1 BARRIERS OF ECO-SUSTAINABLE DEVELOPMENT

not, might reach 12.0 billion or more!

1.1.2 膨胀的需求

一个现代人物质生活与精神生活所需的建筑类型、面积、容积、环境空间、能源、土地等是一个古代人的几十倍乃至几百倍。膨胀的个人需求量乘以膨胀的人口总量是一个庞大的数字,而地球只有一个,由此必然引起能源、土地、供房以及环境危机。

1.1.2 Expanded needs

The needs of building type, area, volume, environmental space, energy and land, for a modern human being to live a material and spiritual life are tens to hundreds times more than those for an ancient. Expanded individual needs multiply by the expanded total population is a large number, while the Earth remains as the same one, will certainly cause the crises of energy, land, housing and environment.

1.1.3 人类必须自控

面对急剧的人口膨胀加剧的能源、土地、住房、环境等危机,治本之道之一,就是人类必须做到自控,尤其是发展中国家如中国、印度。中国和印度两国人口加起来已经比世界人口总量的1/3还多。

著名经济学家(北京大学前校长)马寅初教授早在新中国成立不久就提出了"节制人口"、"优生优质"的先进人口论。但令人愤懑,从1950年代就一直遭到10多年的错误批判。现已清楚地看到由此造成的严重恶果。

1.1.3 Humankind must be self-controlled

Facing rapidly expanding population aggravating the crises of energy, land, housing and environment, one of the fundamental controls is that humankind must be self-controlled, especially in developing countries such as in China and India. The total population of China and India together are now over 1/3 of all the world's total population.

At the early stage in founding new China a famous economist Prof Ma Yinchu(former president of Beijing University) published his advanced theory on population. He suggested "population control" and "high quality of birth and population". Resentfully, since 1950s wrong criticisms have pressed his opinion for more than 10 years. Now we can see clearly that the wrong criticism has brought us a devastating result.

1.1.4 变人口负担为人才资源

我国要变人口负担为人才资源,这的确是一良策。但必须在雄厚的经济基础上大力发展教育特别是高等教育才能做好。这是一项长期的任务。任重道远!

1.1.4 To make population from being a load to be a talent resource

In China to make population from being a burden to be a talent resource is a good strategy. But doing it well must rely on a solid economic base to greatly develop education especially advanced education. This is a long-term task. The responsibility (burden) is heavy and road is long!

1.2 自然灾害

中国每年受自然灾害危害者多于6亿多人次。

1.2.1 大自然灾害

(1) 洪水

全球每年洪水使几亿人受害，几千万间房屋受损，几百亿美元经济损失。中国1998年特大洪水中，3.5亿人受害，死亡4000多人，塌房558万间，损房1205万间。经济损失3000亿元人民币。

1.2 Natural Disasters

In China, every year there is more than 0.6 billion person-time suffered the damages of natural disasters.

1.2.1 Serious natural disasters

(1) Floods

Every year floods throughout the world jeopardize hundreds of millions of people, destroys tens of millions of houses and make tens of billions of USD lost in economy. In 1998, China had a mega-sized flood, in which 0.35 billion person-time was affected, more than 4000 people killed, 5.58 million houses collapsed, 12.05 million houses damaged and 300 billion Yuan RMB lost in economy.

(2) 地震

地震天天有。20世纪以来，发生在中国最惨痛的两次大地震：一次是1976年7月28日唐山8.0级大地震，死亡24.2万人，伤14万多，全城房屋除极少数外几乎全部倒塌，财产损失100多亿元人民币；第二次是2008年5月12日，四川省汶川8.0级大地震，全国有震感，几千万人受害(死亡约7万)。好在得到全国党、政、军、警、民迅速的强力支援和世界人民的支援，使灾区的生产、生活得以迅速恢复。

发生在海底垂直运动的地震还会引发海啸(地震海啸)。2004年12月26日发生在印尼附近印度洋的海啸就是太平洋、印度洋海底岩石板块挤压产生垂直运动的9.3级海底地震引发的。印度尼西亚受害最深，涉及印度、泰国、缅甸、老挝、柬埔寨。死亡30多万人，伤500余万人，房屋倒塌冲入海洋无以计数。

(2) Earthquakes

Earthquakes occur every day. In 20th century in China, two most painful earthquakes have occurred. The first one was on July 28 th 1976, in Tang Shan City, magnitude 8.0, 0.242

million people killed, more than 0.14 million people injured, almost all houses but a few collapsed, 10 billion Yuan RMB lost in property. The second one was on May 12th 2008, in Wenchuan, Sichuan Province, magnitude 8.0, tens of millions people affected (near 70 thousand people killed). Fortunately all had received quick and effective helps from all over China, i.e. government, army, police and people, and the international society. The city has made quick recovery of production and life.

A vertical moving earthquake at seabed might cause a Tsunami. On December 26 2004, in Indian Ocean near Indonesia happened a Tsunami caused by a vertical moving seabed earthquake of 9.3 magnitude. It was the result of the crush between Indian Ocean and Pacific Ocean seabed rocks against each other. Indonesia suffered the worst damage. India, Thailand, Burma, Laos and Cambodia also were affected. More than 0.3 million people killed, more than 5 million people injured. The number of houses collapsed and pushed into the ocean were too much to count.

(3) 厄尔尼诺与拉尼娜

厄尔尼诺是一种影响全球的灾难性气候现象。每隔2~7年，圣诞节前后，南太平洋（正处于夏季）不按通常吹东南信风，而是反常吹南偏西风，把南太平洋暖海水推向东北太平洋。使该处含大量食料的深层冷海水不能上翻，导致大量浮游生物死亡。并连锁反应，使大量鱼群死亡，鸟类死亡，迁徙，肥料减少，农业减产……。与此同时，大气从该温暖的海洋区得到更多高温水蒸气，使气温升高，造成该大气所经地区降雨率减少，形成旱灾。这股高温、高湿气流，由于大气环流及地球自转的影响，流到较冷区，又造成过量降雨，形成洪涝灾害，这就叫厄尔尼诺现象。其间，在澳大利亚的悉尼与阿根廷的布宜诺斯艾利斯（同处南纬34°~35°附近）之间大气气压产生此高彼低、此低彼高的儿童玩跷跷板现象，称南方涛动现象。该现象对厄尔尼诺会起加剧作用，乃合称ENSO现象或事件。

拉尼娜现象则是由于表层海水温度反常地降低，引起的与厄尔尼诺现象方向相反的旱涝灾害，故又称反厄尔尼诺现象。

1998年，40多国旱灾，20多国水灾，中国2009年特大旱灾和2009~2010年冬特大雪灾都与上述气象灾害有关。

(3) El Nino and La Nina

El Nino is a disastrous weather phenomenon affecting the entire planet. Every two to seven years, around the Christmas Day, the South Pacific (just in summer) does not blow normal SE wind. Instead it blows abnormally southwest wind. The wind pushes the South Pacific warm seawater to the northeast pacific. This prevents the cold seawater containing abundant food source from upturning to the sea surface. This results in mass planktons death. In chain reaction, more fish died, more birds died or migrated, fertilizer reduced and agriculture dropped in production… As atmosphere gains more high temperature water vapor from the heated sea water, air temperature rises. It reduces raining chances in the areas where the air passing over and causes droughts there. Affected by circular air movement and the Earth revolving on its own axis, the above air, which contains of high temperature and high humidity, moves to cold region and causes heavy rainfall, flooding and other water logging disasters. This phenomenon is so called El Nino Phenomenon. During El Nino phenomenon, between Sydney, Australia and Buenos Aires, Argentina (the two cities are located near south latitude 34°~35°S), the atmosphere pressure has a phenomenon of that as one falls while another rises as if children play on a seesaw. This is called South Oscillation, it aggravates El Nino and their combination is called ENSO.

La Nina phenomenon is opposite to El Nino. Because the sea water surface temperature anomalously drops down and causes drought or water logging disasters on the direction opposite to El Nino. La Nina is called anti-El Nino phenomenon.

In 1998, more than 40 nations suffered droughts, more than 20 nations were flooded. In China, the 2009 mega-sized drought and 2009-2010 massive snow disaster were both related to the above weather disasters.

(4) 台风、飓风、龙卷风

孟加拉一次强台风,风速72m/s,13万多人死去。美国是多飓风国家:美国佛罗里达州一次飓风将260万km² 地区房屋夷为平地;2005年8月底,新奥尔良市被飓风洪水淹没。中国及其他150多个国家对其紧急援助。我国每年也会受到多次台风袭击,造成沿海地区巨大灾害。

(4) Typhoons, hurricanes, tornadoes (cyclones)

In Bangladesh, a strong typhoon, wind speed at 72m/s, has killed more than 0.13 million people; in America hurricanes occur frequently: in Florida, a hurricane razed all the houses in a region of 2.6 million m² to flat ground; In 2005, at the end of August, a hurricane flood submerged the New Orleans. China and more than 150 countries urgently provided help. Every year China suffers many typhoons' raids causing serious damages in coastal areas.

1.2.2 其他自然灾害

雷击(温度可达25000℃)、森林火灾、冰雹、火山喷发、滑坡、沙漠化以及流行病如"非典"、禽流感等等造成的负效环境,也是生态可持续发展的巨大障碍。

1.2.2 Other natural disasters

Being strucked by lightning (temperature may reach 25000℃), forest fires, hailstones, volcanic eruptions, landslides, desert spreads and epidemics such as "SARS", bird influenza, and etc. are all categorized as negative environmental effects. They are also big obstacles to eco-sustainable development.

1.3 人类负效行为

1.3.1 人类的严重负效行为

(1) 破坏绿色世界,加剧大气温室效应与气候变暖

大气中的二氧化碳(CO_2)有一种特性:能透过短波辐射,如太阳辐射;同时吸收长波辐射，如地面辐射；使大气和一定深度的陆地与海水升温，这就是通称的大气温室效应（AGE）。其累积效果会使气候过度变暖。

远古时代，火山喷发带出大量 CO_2 形成大气温室效应，使气候变暖，是产生生命的必要条件之一。否则，地面将一直处于寒冰状态（-50℃以下）无生命产生。不幸的是，近代人类的生产、生活以及动植物的呼吸作用，已经并仍在排出大量 CO_2；同时，又已经并仍在严重破坏绿色世界，使光合作用吸收 CO_2 的力量大大减弱，严重加剧了大气温室效应与气候变暖，最终带来毁灭性灾害：群山冰雪消融、河川断流、湖泊干涸、地表淡水断源、两极冰山融化，海水升高可能达 60m，全球毁灭性大灾难！

人类破坏绿色世界最严重的行为是建筑侵占绿地与滥伐森林，这不仅会加剧大气温室效应与气候变暖，还会加剧水土流失、洪水灾害与生态失衡。

建筑是破坏绿色世界的最大侵犯户，也是还自然绿色世界的最强潜力者。

大气温室气体已发现有 30 多种。CO_2 温室效应最强，且与建筑关系最密切，故论及较详。

1.3 Human Negative Behaviours

1.3.1 Man's serious negative behaviours

(1) Damaging vegetation coverage, aggravating atmospheric greenhouse effect and climate warming

The carbon dioxide (CO_2) in air has a character of transmitting short wave radiation such as solar radiation while absorbing long wave radiation such as ground radiation. This can heat up the temperatures of air and of certain depth of the land and seawater. This is so called "Atmospheric Greenhouse Effect (AGE)". Its accumulative effect can cause climate over warming (COW).

In remote ancient times, the volcanic eruptions blew out vast amount of CO_2, which formed the AGE causing climate warming up. That is one of the essential factors for the origin of lives. The ground would have stayed in the cold ice state (below -50℃) without it. No life would have been born. Unfortunately, in modern times, the man's production and living and the breathing of animals and plants have discharged and is still discharging vast amount of CO_2 while seriously damaged and is still damaging the green environment. It has greatly weakened the green force absorbing CO_2 under photosynthesis. Consequently, this is aggravating the AGE and the COW. Eventually, it will cause exterminative disasters: ices and snows on mountains melted, rivers stopped flow, lakes dried out, ground fresh water sources stopped and the icebergs in the two poles melted, that sea level might rise 60 meters, causing exterminative disaster throughout the world!

Man's most serious damaging behavior to the green environment is occupying the forest by buildings and deforestation. This can not only aggravates AGE and COW but also aggravates soil erosion, floods and eco-conservation imbalance.

Construction activity is the biggest invader damaging green environment while remains potentially most powerful one to return the nature a green world.

1.3 人类负效行为
1.3 Human Negative Behaviours

Among more than 30 types of the atmospheric greenhouse gases discovered, CO_2 has strongest greenhouse effect and is related to buildings the most. Hence we will discuss it in details.

(2) 破坏臭氧层

在离地面25km以上的大气层中,有一层厚约30km的臭氧层,是保护地球及地球生物免遭太阳紫外线伤害的保护伞。由于太阳紫外线的照射,使大气中的氧分子(O_2)分解成氧原子(O)。在碰撞中,三个氧原子,或一个氧分子与一个氧原子结合在一起就形成了一个臭氧分子(O_3)。众多的臭氧分子形成了臭氧层。正是这层由太阳紫外线造成的臭氧层,吸收了射向地球99%的紫外线,使地球及其生物免遭伤害。剩余1%的太阳紫外线,到达地面,杀菌、消毒、促进维生素D的转成,促进骨组织钙化及骨骼生长,帮助儿童预防佝偻病。

但长期以来人们使用的空调器、冰箱、汽车、计算机以及发泡剂、清洗剂等含有大量氯氟烃类物质。氯氟烃分子被排放到大气中,在太阳紫外线作用下会裂解,放出游离的氯原子。氯原子很活跃,它会从臭氧分子"O_3"中拉出氧原子"O",臭氧分子就减少。一个氯原子大约可破坏10万个臭氧分子,使臭氧大量减少,形成"空洞"(臭氧浓度比正常少了75%)。地球南、北极均已发现臭氧空洞,南极臭氧空洞相当于三个中国陆地之和(2800多万平方公里),并以每年相当于一个中国或美国陆地面积的速度扩展。许多企业虽已有改进措施,但长期积存在大气中的危害气体还会起破坏作用。

(2) Damaging ozonosphere

At 25km above ground, the atmosphere contains an ozonosphere of about 30km thick. It's the umbrella protecting the Earth and its living beings from the damage of solar ultraviolet rays. Solar ultraviolet rays can break down the oxygen molecules (O_2) into oxygen atoms (O). During impact, three oxygen atoms or one oxygen atom and one oxygen molecule integrate into one ozone molecule (O_3). Lots and lots of ozone molecules form the ozonosphere. It is this ozonosphere formed by solar ultraviolet rays that have absorbed 99% of the solar ultraviolet rays towards the Earth so that the earth and its living beings are protected from the rays' damage. The remaining 1% ultraviolet rays reached the ground. It can kill germs, disinfect air and help the formation of vitamin D, which calcifies the bones, encourages their growing and protects children from rickets.

But over the years, people have been using the air conditioners, refrigerators, automobiles, computers and chemicals such as vesicatory (vesicant), abstergent, and etc. These contain large amount of hydrochlorinefluor. When emitted to the atmosphere, under solar ultraviolet rays' action, hydrochlorinefluor molecules can be split and release the dissociated chlorine atoms. They're very active and can draw oxygen atom (O) from ozone molecule (O_3) and reduce ozone molecules. One chlorine atom can break down about 0.1 million ozone molecules and make ozone less and less. This process eventually resulted the formation of "ozone hole" (in which ozone consistency is 75% less than the normal consistency). In the South Pole and North Pole, ozone holes have been found. In South Pole the ozone hole's area is as large as the size of three China combined (more than 28.00 million km^2). Every year it is still spreading with an area like one Chinese land or one American land. Although many enterprises have made improvements, the accumulated harmful gas, over a long period, in the atmosphere is still damaging the air.

(3) 人为水污染

在中国，生产、生活排放的污水已使全国80%的地表水与地下水受到污染。

(3) Water pollution by man

In China, 80% of surface water and groundwater have been polluted by the wastewater from manufacturing and living.

(4) 人为大气污染

中国每年生产、运输、2.6亿多户的炊事以及人、动物、植物的呼吸作用排放大量二氧化碳和其他烟尘已使大气受到严重污染。全国城市中仅1/3人口可吸到新鲜空气。CO_2是加剧大气温室效应与气候变暖最主要的气体。中国和美国是全球最大排放户。

(4) Air polluted by man

Every year, in China, the air has been seriously polluted by production, traffic, and cooking from more than 0.26 billion homes' and other smoke and dust. Even the breathing of people, animals and plants emits vast amount of CO_2 into the atmosphere. In cities only one third of people can breathe in fresh air. CO_2 is the main source aggravating atmospheric greenhouse effect and climate warming. China and US are the global top emitters.

(5) 人为火灾

中国每年人为火灾均在10万次以上。2006年达20多万起。2003年11月13日，湖南衡阳一起火灾，除平民伤亡外，仅消防队员就有20位英勇殉职。每年全国火灾中经济损失都在300亿元人民币以上。

(5) Fires by man

In China more than 0.1 million cases of fires occur each year. In 2006, more than 200 thousand fires happened. The most grievous one occurred on November 13th 2003 in Hengyang City, Hunan Province. Among the killed and injured, 20 firemen heroically died at their posts. In fires each year more than 30 billion Yuan RMB lost in economy in China.

(6) 人为生产事故

由于管理者的失职或工人的错误操作，近几年来每年中国都发生多起生产事故，特别是在开采业中的瓦斯爆炸、地下水涌灌作业隧道、井喷等。工人遭受惨重伤亡，国家经济受到巨大损失。

(6) Industrial accidents by man

In China during the past few years, a lot of industrial accidents occurred due to managers' negligence or workers' wrong operation, especially in mining sector: such as gas explosion, ground water pushed into working tunnels, blowouts, and etc. Workers suffered heavy casualties and national economy suffered tremendous lost.

(7) 核害

人们绝难忘记1945年日本广岛、长崎两城市被两颗原子弹夷为平地，30万人惨遭杀害！更多的人受到核辐射伤害！人们也绝难忘记前苏联契尔诺贝利核电站1984年4月的核泄漏悲剧，14万人严重受到核辐射伤害，大范围生态受到严重破坏。

1.3 人类负效行为
1.3 Human Negative Behaviours

2011年3月11日,日本发生9.0级地震,引发浪高10m的巨大海啸,致使福岛核电站严重破损,产生严重核泄漏灾害!

长期的军备竞赛,我们的地球已经遭受了几百次核爆炸污染!21世纪,核害加剧的隐患一直存在。

(7) Nuclear disasters

People can never forget that in 1945, the two cities, Fukuyama and Nagasaki of Japan were leveled to the ground by two nuclear bombs and 0.3 million people miserably killed! Much more people injured (harmed) by nuclear radiation. People also remember clearly that in former CCCP there happened a tragedy of nuclear leak at Chernobyl nuclear power station, in April 1984, in which 0.14 million people were seriously injured by nuclear radiation. The eco-environment in a large area was badly damaged.

On March 11, 2011, an earthquake of 9 magnitude happened in Japan causing a huge Tsunami, whose wave reached 10m high. Consequently, the severe damage to Fukushima nuclear power station has led to a devastating nuclear leak calamity!

During the long time armament rivalry, the earth has suffered the pollutions of nuclear blasts for several hundred times! In the 21st century, the nuclear pollution might be aggravated due to the hidden danger exists all along.

(8) 战争

自古以来,特别是冷兵器变为热兵器以来,战争是屠杀百姓、侵害人民财产、毁坏建筑及其环境、危害生态可持续发展的最大祸首。

(8) Wars

Since ancient times, especially since cold weapon became hot weapon, war has been the chief culprit in killing people, encroaching properties, destroying buildings and environment and damaging eco-sustainable development.

(9) 恐怖分子危害

2001年9月11日,美国纽约世贸中心双塔超高层建筑被恐怖分子控制的两架喷气机先后撞击起火,两小时内相继倒塌,两千多人丧生。2005年7月英国伦敦,2008年底和2009年初,印度和巴基斯坦又数次遭恐怖分子爆炸袭击。2009年7月5日中国乌鲁木齐遭受恐怖分子烧、杀、打、砸、抢的袭击。

恐怖主义是对世界和平与可持续发展最严重危害之一。

(9) Terrorist damages

On September 11th 2001, the two WTCT (World Trade Center Tower) super high-rise towers were impacted one after another by two jets controlled by terrorists. Soon the buildings were burning and in two hours collapsed. More than 2,000 people were killed. In July 2005 in London UK, and at the end of 2008 in India and in the beginning of 2009 in Pakistan, several blasts by terrorists have been reported. On July 5th 2009 the Urumqi area in China suffered terrorist's heavy attack of killing, beating, breaking in, robbing and firing.

The terrorism is one of the worst threats to world peace and sustainable development.

(10) "文化大革命"浩劫

中国1966~1977年的"文化大革命"是一场空前的浩劫，对政治、经济、科学、艺术、文化等所有领域的生态可持续发展都是一个严重负效应。

(10) "The Great Cultural Revolution" calamity

"The Great Cultural Revolution" was an unprecedented calamity provoked in China from 1966 to 1977. It had seriously negative effects on the eco-sustainable developments of all areas of politics, economy, science, arts and culture and etc.

1.3.2 其他人为危害

吸毒在全球成亿计地夺去人的躯体、意志、家庭。有些吸毒者还会进行盗、骗、抢、杀等犯罪行为。

艾滋病在全球正以每天17000人的速率传染。我国已发现8种类型艾滋病感染者，人数已超过800万，多为青少年。对艾滋病患者，应以关爱之心相待，以助康复。

其他性病我国已有600多万人感染。

贪污腐化自古以来就是对为官担职人群中意志薄弱者的恶性污染。中国建筑业是贪污腐化最严重户之一。

上述人为负效行为已使社会环境、经济环境、政治环境、精神环境受到严重污染。

1.3.2 Other harms by man

Drug abuse (taking) has deprived hundreds of million people from their health, volition and families throughout the world. Some drug addicts involved in crimes such as stealing, deceiving, robbing and killing.

AIDS has been infecting sufferers with the rate of 17000 persons per day in the world. China has found sufferers of eight types of AIDS. The total number of sufferers has reached over eight million. Most of them are youngsters. We should treat them with kindness and love to help them gain recovery.

The number of sufferers of other venereal diseases (V.D) has also reached over six million in China.

Corruption since ancient times has always been a vicious pollution to the weak-minded officers. In construction industry corruption is the most serious one.

The above man's harmful behaviours have seriously polluted the social, economical, political and psychological environments.

2 温故知今与创新——中国建筑回顾
2 LOOKING BACK TO THE PAST TO UNDERSTAND TODAY AND CREATE THE NEW——Reviewing the Chinese Architecture

本章为促进建筑的生态可持续发展回顾中国建筑。

The review of the Chinese architecture in this chapter is for the promotion of eco-sustainable architectural development.

2.1 乡土建筑

2.1.1 概况

我国是一个拥有56个民族的大家庭。建筑中科技原理的同一性与民族文脉的个异性的结合已经并正在构建中华建筑的辉煌灿烂史。

不同地区的建筑适应着各地区的气候和地理状况。众所周知，我国幅员辽阔，地形、气候复杂多变，平原只占国土面积的12%，各种沟、洼、坡地占70%以上。从东北、西北、台湾、海南至南沙群岛等，寒、温、热区都有。仅西部地区，有海拔4000m以上的青藏高原，也有低于海拔155m的、全国最低的吐鲁番盆地。该盆地素有"火州"之称。1998年夏，在该盆地丝绸之路—古城堡（图2-1），作者夏云测得：气温40℃，地面温度高达66℃。据报，2006年7月末，该地区有一处气温高达55℃（大气温室效应影响）。而在云南大理，全年室外气温≥30℃的天数，每年平均只有1.6天。

我国乡土建筑乃世界上多样性最多的建筑，如：云南傣族的竹楼（彩图2-2），陕、甘、宁、青、豫的窑洞（彩图2-3），吐鲁番的葡萄乡里，海南的椰林别墅，广西漓江的山居，云南丽江的美景，苏、杭的园林与小桥流水人家，以及其他园林湖榭等等（彩图2-4～彩图2-6为选例）。没有千万不同的个性也就没有绚丽多彩的世界。

2.1 Vernacular Architecture

2.1.1 General

China is a great family of 56 nations. In buildings the combination of the identity of science principles and the individualities of different cultural context has been forming the splendid history of Chinese architecture.

2 LOOKING BACK TO THE PAST TO UNDERSTAND TODAY AND CREATE THE NEW——Reviewing the Chinese Architecture

In different geographical regions, buildings adapted to their own climate and location. China is well known of its vast territory, complicated geography and varied climate. Besides the plain taking 12% of the total territory, more than 70% of the country consists of various gullies, depressions and sloping fields. From the Northeast, Northwest to Taiwan, Hainan and Nansha Islands, the climate changes from frigid, temperate to torrid. Only take an example to look at part of the West: Qinghai-Tibet Plateau, is more than 4,000 meters above sea level whereas Turpan Basin in Xinjiang is 155 metres below sea level. Turpan is very hot in summer hence be called "Fire land". In 1998 summer, at an ancient earthen castle on the Silk Road of the basin (Fig. 2-1), the author Xia Yun has measured the air temperature 40℃ while the ground temperature reached as high as 66℃. It was reported, at the end of July 2006, there was a recorded air temperature at 55℃ in this region (influenced by the atmospheric greenhouse effect). In contrast in Dali City Yunnan Province, the total average number of days of outdoor air temperature reaching above 30℃ is only 1.6 days per year.

Chinese vernacular architecture has the most varieties in the world. For examples, Yunnan Dai group's bamboo houses (color picture 2-2), Cave dwellings in Shaanxi, Gansu, Ningxia, Qinghai, Henan provinces (color picture 2-3), Turpan grape villages, Hainan villas among coconut trees and Guangxi hillside dwellings by Lijiang River. The wonderful Yunnan Lijiang riverside landscapes, Suzhou and Hangzhou Gardens and other type of houses by streams through small bridges and numerous gardens, lakes, pavilions, etc. (color pictures 2-4~2-6) are all good examples. There is no such a colourful world without thousands upon thousands of individualities.

图 2-1 丝绸之路一座古城堡，1998 年夏，
作者曾测到气温 40℃，地面 66℃

Fig. 2-1 An ancient earthen castle on the Silk Road, in 1998 summer, the author has measured air temperature of 40℃, ground temperature of 66℃

2.1.2 乡土建筑的优点

概言之：

2.1 乡土建筑

2.1 Vernacular Architecture

(1) 顺天地之道
即适应当地气候、地质建房。
(2) 承祖传科艺
中国各民族建筑历史相沿的建筑技术与艺术，都是代代相传递、发展的。
(3) 扬民族文脉
中国乡土建筑饱含各民族科技、艺术和文化，它们文脉世代相传，并发扬光大。

2.1.2 Advantages of vernacular architecture.

In summary：
(1) According to the local conditions of climate and geology
People adapt the construction to the local climate and geology；
(2) Inheriting ancestors' science and art
Historically, building technology and art in every Chinese nation are inherited and developed from generation to generation；
(3) Preserving national heritage
Chinese vernacular architecture fully contains local context of science, technology, art and culture of each nation. They have been inherited and carried forward from generation to generation.

2.1.3 乡土建筑的缺点

我国乡土建筑耗能用地方面的缺点可用 10 个汉字概括：分散、层少、耗能多、占地大。乡土建筑分散、层少是耗能多、占地大两大原因。图 2-7 示例为陕西某一有名富村，住房都是两层，人均建筑面积≥30m² (为当时五倍城市居民人均建筑面积)。人均建房占地约 55m²。乡土建筑绝热差也是耗能多的主因之一。我们曾进行冬、夏温度测量：图 2-8 是受测房屋之一，冬季室外气温－2℃（并不很冷），室内呼吸线气温却只有－0.1℃，棉垫下热炕上面才有 18℃（冬季每晚烧炕）；夏季室外气温 34℃时室内气温竟高达 32℃！又如西安近郊一建筑，24 砖外墙砖混二层楼房，室外气温 38℃时，室内气温竟高达 34℃！什么原因呢？原因是：

1) 绝热差：缝隙透气大；门窗接缝不严；墙顶与檐口交接处空隙大；外墙与屋顶热阻不够。

2) 室内弃热多，又不利用：炊事、烧炕两项中 85％以上热能变为飞灰，污染大气与环境。

2.1.3 Disadvantages of vernacular architecture

In terms of energy-consumption and land-use, Chinese vernacular architecture has shortcomings. In short, it is scattered, low-story and it consumes more energy and occupies more land. Vernacular buildings' scattering planning, low-story characters are the two main reasons of its high energy consumption and land occupation. Fig. 2-7 shows an example in a rich village in Shaanxi. All houses are two-story buildings with average floor area above 30m² per capita (five times that of a urban citizen then). The land occupied per capita is about 55m². Poor insulation is also one of the main reasons causing higher energy consumption. We have done an experiment comparing temperatures measured in winter and summer in different buildings. Fig. 2-8 is one of the houses measured. When outdoor air was negative 2℃ (not very cold), indoor air at breath line level was only negative 0.1℃. On the warm Kang's surface beneath the cotton mat, it was 18℃ (in winter every night the Kang must be heated). In summer when outdoor air was 34℃, indoor air has reached 32℃! Another example in one of Xi'an's suburbs is a two-story concrete-brick house with 24cm thick clay brick outer wall. When outdoor air was 38℃, indoor air has reached 34℃! What's the reason? The reasons are:

1) Poor insulation: high infiltration and poor air tightness of doors and windows; big gap between wall top and eaves; not enough thermal resistance of exterior walls and roof;

2) Lacking of recylcle use of abandoned heat internally: in cooking and heating the Kang, more than 85% of heat becomes smoke polluting air and environment.

图 2-7 陕西某富村住宅区：分散、层少、耗能多、占地大
Fig. 2-7 Shaanxi, a rich village's settlement: scattered, Low-story, consumed higher energy and occupied more land

图 2-8 绝热调查农房之一
Fig. 2-8 One of the village houses for heat insulation investigation

2.2 城市建筑

2.2.1 引言

经济大发展必然引起城市大发展，城市大发展建筑及其环境必先行。因为任何领域的发展必先由建筑为其提供生产、科研、生活、管理用建筑空间。现代城市建筑的特点是多层、高层、超高层。

2.2 Urban Architecture

2.2.1 Introduction

Great development of economy certainly leads to great development of cities, and in

2.2 城市建筑
2.2 Urban Architecture

great development of cities, building and its environment should be the first to be developed. Because for any other field's development, construction industry must supply building space for the production, research, living and management. The characteristics of today's urban buildings are multiple-story, high-rise and super high-rise.

2.2.2 城市建筑优点

(1) 与分散、层少的乡村建筑比，城市建筑节约资源：同样建筑面积，城市建筑层多，地基少，屋面少，室内外热交换面少，有节能、节地、节材、节工、节时、节财和减排的好处。

(2) 城市建筑每平方米占地面积产值高。

2.2.2 Advantages of urban architecture

(1) Comparing to the scattered, low-story building in villages, the urban building can save resources. With the same floor area urban building has multiple stories, less foundation, less roof, less building external envelope for heat exchange. This benefits saving energy, land, material, work, time and money and reduces discharge of wastes.

(2) Every square metre ground area occupied by the urban building has higher output value than the building in villages.

2.2.3 城市建筑缺点

(1) 冬夏绝热差，耗能多。
(2) 侵占绿地多，加剧大气温室效应、夏季热岛负效应与用电高峰。
(3) 防震、防火能力低，紧急疏散难。
(4) 房高招风(实验：房高10m处风速6m/s；房高30m处风速9m/s；房高90m处风速15m/s)，故高房会加剧空气渗透与对流换热，使冬失热、夏得热增多，耗能加大。
(5) 高层建筑与超高层建筑有房颤。
(6) 玻璃幕墙特别是镜面玻璃幕墙有光、热污染，危害环境。

2.2.3 Disadvantages of urban architecture

(1) Poor insulation in winter and summer hence high energy consumption.

(2) Occupying large green area, aggravating AGE, having negative thermal island effect in summer and increasing electricity peak load.

(3) Low abilities in earthquake proofing, fireproofing and having difficulty in urgent escape.

(4) Tall building catches the wind (according to experiments: wind speed (w. s.) reaches 6m/s at the building height (b. h.) of 10m, w. s. 9m/s at b. h. 30m and w. s. 15m/s at b. h. 90m). So tall building aggravates air penetration and convective heat exchange. It also increases winter heat loss, summer heat gain and energy consumption.

(5) Building vibration exists in high-rise buildings and super high-rise buildings.

(6) Glass curtain wall especially mirrored curtain wall have light and heat pollutions damaging environment.

城市建筑例

彩图2-9为20世纪80年代我国某大城市一角，绿化少，绝热差，热岛负效应严重。现在大有改进，被有关国际组织选为100万人以上"花园城市"，她就是深圳市。彩图2-10为巴黎一镜面幕墙高层建筑，两侧36层镜面幕墙支撑一座6层建筑为屋顶。使用者（办公人员）说："我们像住在盒子里一样，视觉和心里感受与外界信息隔绝，常有忧郁感"。有的法国教授、建筑师认为，该房反射大量太阳能造成光热污染，同时室内又需长年空调维持好微气候，实是一大缺点。

Urban building examples

Color picture 2-9 shows a section of a Chinese mega city in 1980s. It had less greening, poor insulation, and serious negative thermal island effect. Now it has been greatly improved and chosen by an international organization as a "Garden City" of over one million population. She is Shen Zhen City. Color picture 2-10 shows a mirrored curtain wall high-rise building in Paris. Both sides' of the 36-story mirrored curtain wall high-rise buildings support a 6-story building as the roof. The users (employees) said: "It seems that we are living in a box. The views are blocked and we feel psychologically isolated from the outside world. We often feel heavy-hearted". Some France professors and architects think that this building reflects too much solar energy causing light and heat pollution. It also needs indoor air conditioning to keep the microclimate comfortable all year round. This is a serious shortcoming.

思考

（1）有医学专家指出：从生理反应与心理感受看，绝大多数人并不喜欢在高层或超高层建筑内工作、生活，你意如何？

（2）高层和超高层建筑未来如何？

（3）当代城乡建筑在重大天灾人祸下仍很脆弱，并成为杀手，怎么办？

Thinking deeply

(1) Some medical experts have found that from study of the physiological reaction and psychological feeling, most people don't like to work and live in high rise building especially super high-rise building, what's your opinion?

(2) What is the future of high-rise buildings and super high-rise buildings?

(3) Under serious natural and man-made calamities today, urban & rural buildings are very fragile and become killers. What to do next?

2.3 新概念建筑

2.3.1 引言

我国从20世纪70年代起，为克服上述缺点，有关单位：建设部、农业部、大专院

2.3 新概念建筑
2.3 New Concept Architecture

校、研究所、设计院……他们已开始新概念建筑探索，其历程可概括为：传统建筑→节能建筑、太阳房→节能、节地建筑→绿色建筑、生态可持续建筑。

2.3 New Concept Architecture

2.3.1 Introduction

To overcome the above shortcomings, since1970s the relative Chinese organisations: the Ministry of Construction, the Ministry of Agriculture, universities, colleges, research institutes and design institutes and etc., have started to explore a new concept architecture. The process may be summarized as: Conventional buildings→Energy saving buildings and solar buildings→Energy and land saving buildings→Green architecture and Eco-sustainable architecture.

2.3.2 能源危机引发了节能建筑与太阳房的研究与推广

(1) 能源危机

20世纪70年代，由于政治和经济的原因引起售油国与购油国（主要是发达国家）的矛盾，导致世界性能源危机。我国也同样存在此危机，主要原因是：百业待兴，供不应求，耗能甚大，浪费惊人。这表明我国急需开展"开源节流和减少污质排放"的科研与实践。建筑领域是用能排污最大户之一，"开源节流、减排污质"任务甚艰巨，但潜力也十分巨大。

2.3.2 Energy crisis led to the research and implementation of energy saving buildings and solar buildings

(1) Energy crisis

In 1970s, political and economic issues have caused the conflicts between the oil-saling countries and oil-buying countries (most are developed ones) and resulted in the world energy crisis. China also has energy crisis. This is due to thousands of incomplete business, exceeding demands beyond supplies, high energy consumption, and shocking wastage. This indicates that China must urgently develop the research and practice of "developing energy resources, energy saving and reducing pollutants' emissions". Construction Industry is one of the top energy users and emitters. Its duty of "developing energy resources, energy saving and reducing emissions" is very heavy but with great potential.

(2) 节能含义及原则

能源危机惊起全球重视节能以来，对节能含义有两个阶段理解：初始阶段狭义的理解，继而上升到广义的理解。

(2) The meaning and principle of energy saving

Since the energy crisis has alarmed the world to focus on energy saving, we have passed two stages to understand the meaning of energy saving. First, from narrow sense, then, from broad sense.

狭义的"节能"指的是节约常规能源：煤、气、油、柴、电，用节能专业术语"节能率"来衡量，例如，冬季某室内保持20℃，每天耗煤100kg，通过改进，用太阳能并加强保温，只耗煤50kg即可保持室内20℃，那么节能率就是50%，即：

$$节能率 = \frac{原用能 - 改进后用能}{原用能} \times 100\%$$

The narrow sense of energy saving is to save conventional energy sources: coal, gas, oil, firewood and electricity. The terminology: Energy Saving Rate (ESR) is used to estimate the energy saving. For example, to keep a room air at 20℃ in winter, the coal's consumption is 100kg per day. Now through an improvement of using solar energy and increasing thermal insulation, to keep the room at 20℃, it only needs 50kg coal per day. So the ESR is 50% i. e.

$$ESR = \frac{(Original\ energy\ consumed) - (Energy\ consumed\ after\ improving)}{Original\ energy\ consumed} \times 100\%$$

广义"节能"含义有二：开发利用可持续能源；有效用能。开发利用可持续能源指的是开发利用太阳能、沼气等。有效用能指的是一切用能都应提高用能效率，即提高最终产品有效含能量与输入总能量之比：

$$用能效率 = \frac{最终产品有效含能量}{输入总能量} \times 100\%$$

From broad sense, energy saving has two meanings: to develop and use sustainable energy sources and be energy efficient. Developing and using sustainable energy means to develop solar energy and marsh gas (firedamp; methane), etc. Energy efficiency means that to increase efficiency in all stages of energy-use i. e. to increase the rate of efficient energy content in the final production to the total energy input.

Energy Efficiency Rate (EER):

$$EER = \frac{Efficient\ energy\ in\ final\ production}{Total\ energy\ input} \times 100\%$$

思考：节能率与用能效率能达到100%吗?

Thinking deeply: Can ESR and EER reach 100%?

例如，通过玻璃窗直接得热的太阳房，节能率可达60%～70%，但对投到玻璃外表面的太阳能来说，用能效率只有30%左右，即投到玻璃外表面的太阳能中只有30%转化为室内供暖的热能。又如：用太阳能电池夜间照明取代传统电照明，对节约传统电来说，节能率是100%。但迄今国内外太阳能电池将太阳能转化成电能的效率一般只是16%～20%，用能效率仍很低。

For example, if a solar house's direct heat gain is from glass windows, its ESR may reach 60%～70%. But for the solar energy reached the external surface of the glass, the EER might be only 30%, i. e. only 30% of the total solar energy reached the external surface of the glass can be converted into heat to heat up the room. Another example: comparing using solar cells for night lighting to conventional electrical lighting, the ESR of saving the conventional electricity is 100%. But the EER of solar cell to convert solar energy into electricity is only 16-20% home and abroad till now. So the efficiency of energy use is still very low.

2.3 新概念建筑
2.3 New Concept Architecture

能量守恒定律

一切用能都要受到能量守恒定律的制约。该定律扼要地说就是：

① 能量既不能被创造，也不能被消灭，它可以从一种形式转变到另一种形式，但宇宙总能量恒定不变。

② 能量形式转变过程必有能量损失。

澳大利亚一位资深科学家 Greenland 曾通俗地说明上述能量定律：第一句话是："You can't get any thing for nothing"，直译："你没有东西就不能得到任何东西"。我国早有类似的谚语："巧妇难为无米之炊"；"欲收必种"；"欲得必付"等。他的第二句话："You can't break even"，"不可能没有盈亏"，"有得必有失"甚至"得不偿失"。我国还有"事倍功半"，"事半功倍"的古谚语描述做事效率高低的说法，其积极意义都是鼓励人们争取高效率。有效用能另一含义就是尽快开发利用干净、安全、永久性的可持续能源取代传统的污染性、不安全、有限性的非可持续能源。

The law of conservation of energy

The law of conservation of energy governs all energy use. The law may be summarized as:

① Energy can neither be created nor destroyed. It may be converted from one form to another, but the total amount of energy in the universe remains constant.

② Energy loss always follows energy conversion.

An Australian senior scientist Mr. Greenland has explained the above energy law in plain English. First: "You can't get anything for nothing". In China there were similar ancient proverbs such as: "A good housewife can't cook a meal without rice", "To harvest one must plant"; "To gain one must pay", etc. Greenland further explained it as: "You can't break even". It means that "Loss follows gain" or even "The loss outweighs the gain". Similarly Chinese ancient proverbs say: "Receive (get) half the result with twice the effort", "Receive (get) twice the result with half the effort". These proverbs are used to estimate high or low efficiency in doing things. Its positive meaning is to encourage people to win over high efficiency. Another meaning of energy efficiency is to develop and use the clean, safe, and eternal sustainable energy instead of the conventional pollutant, unsafe and unsustainable energy as early as possible.

节能的总原则是：开源节流。即开发可持续能源，减少能量损失和浪费。

The general principle of energy saving is "to increase income while decrease expenditure" i. e. to develop sustainable energy while reduce energy loss and wastes.

(3) 中国节能建筑概况

我国已兴建大批太阳能住宅、学校、办公楼、掩土太阳房、农用温室等。设计理论（含CAD）、新材料（相变贮热、热镜玻璃、无滴水温室塑料膜、透明太阳能电池板等）的研究、制造和应用均已取得可喜的成绩。太阳能热水器的生产和应用已名列世界前茅（但离全国2亿5千万户普遍应用还差很远），太阳能照明、太阳灶、太阳能游泳池也得到较大发展，多项成果已获国家奖、专利权或国际奖。

(3) General picture of Chinese energy saving buildings

In China many solar dwellings, schools, office buildings, earth sheltered solar houses and agricultural green houses have been built. Design theory (incl CAD), and the study, manufacture and use of new materials (phase change heat storage; heat mirror glasses; no drop plastic film in green house; transparent solar cell panel, etc.) have great achievements. Production and use of solar water heaters have topped the list in the world although there is a long distance to reach its wide application in all Chinese 0.25 billion families. Solar lightings, solar cookers and solar swimming pools have also been developed well. Many research outcomes have won national awards, patent rights or international awards.

中国发展节能建筑初期，从技术处理上有两类：一类是加强围护结构绝热为核心技术；另一类是利用太阳能为主的太阳房。现已达成共识：节能建筑必需加强绝热，同时加强综合用能。综合用能，就全国城乡来说，应以太阳能与沼气为主。此外，各地根据当地条件还可采用风能、水力能、地热能以及室内弃热，或其他优化组合用能。

In early stage of developing energy saving buildings in China, there were two technical methods: One focused on strengthening envelopes' thermal insulation; another on the utilization of solar energy i.e. solar buildings. Now it is commonly believed that energy saving buildings should strengthen the envelopes' thermal insulation as well as comprehensively use energy. Comprehensive use of energy sources means that: all Chinese cities and villages should mainly use solar energy and methane. Furthermore, each area might use wind energy, hydropower energy, geothermal energy and indoor abandoned heat or other optimal combinations of energy according to the local conditions.

2.3.3 绿色与土地危机引发了节能节地建筑

(1) 引言

20世纪80年代起，绿色与土地危机的进一步加剧，揭示出节能建筑仍存在和传统建筑一样占地多、侵占绿地多的缺点。我国东北、华北、华中、西北以及西南兴建的大批太阳房，如前所述，虽节能率可达60%～70%的好效果，但占地仍很多。当时大多数人、设计人员、科研人员较普遍缺乏土地危机意识。用户过多要求建筑面积，设计者也未考虑节地，致使太阳房用户人均建筑面积仍在 $30m^2$ 以上，人均用地面积仍在 $50m^2$ 以上。全国城乡每年新建房超过10亿 m^2 建筑面积，大量绿地与耕地被占。

建筑是侵占绿地与耕地的最大户。

2.3.3 Green and land crises led to energy and land saving buildings

(1) Introduction

Since 1980s, the crises of green and land have been further aggravated. This revealed that the energy saving buildings have the shortcomings of occupying large land and green area the same as the conventional buildings. In Chinese Northeast, North, Centre, Northwest and Southwest regions, a lot of solar buildings have been built, as mentioned above. The ESR of these buildings has got a good result of 60%～70%. But the land occupation rate was still high. At that time, the ordinary users, designers and researchers generally

2.3 新概念建筑
2.3 New Concept Architecture

were ignorant of the green and land crises. Building users wanted more building floor area while the designers still didn't consider saving land. As the result, in solar buildings the average floor area per capita exceeded 30m^2, land-use area exceeded 50m^2 per capita. Every year the new constructions throughout all Chinese cities and villages are over 1.0 billion m^2 floor area. Vast areas of green and farmland have been occupied.

Buildings are the biggest intruders occupying green and farmland.

(2) 节能节地建筑的产生

建筑不仅是侵占绿地与农田的最大户,也是耗能与浪费能量的最大户之一。

20世纪80年代后期,我国建筑研究者(含本书作者)首先提出:"我国节能建筑与国外不同,必须同时具有节能节地的属性才符合我国国情,才有推广价值。"此后,乃陆续出现了节能节地建筑的研究、实验与学术交流。西安建筑科技大学等还开设了本科生与研究生的"节能节地建筑"课。全国有关单位的研究、实践,取得了相当成果,有的已获国家及国际奖。

(2) Emergence of energy and land saving buildings

Buildings are not only the biggest intruder occupying green area and farmland but also one of the biggest ones consuming and wasting energy.

In the late 1980s, Chinese building researchers (incl the authors of the book) first suggested that: "Chinese energy saving buildings are different from the foreign ones. They must have the character of both energy saving and land saving at the same time so as to meet the Chinese conditions and be worthwhile implementing." Since then researches, experiments and academic exchanges about energy and land saving buildings have emerged one after another. Xi'an University of Architecture and Technology and some other universities have included the subject of Energy and Land Saving Architecture into the undergraduates and postgraduates' curriculum. The researches and practices by relevant organisations have made certain achievements. Some of them have won national or international awards.

(3) 节能节地建筑含义

节能节地建筑决不意味着要人们在低标准的狭窄空间里生活,而是预示着人们将不断以新的科技手段,充分利用洁净、安全、永久性的太阳能及其他可持续能源,取代传统污染的、不安全的、有限性的非可持续能源,并以美观的形象、适宜的密度、地上地下、海上、陆地相结合的建筑为人们创造美好的生活空间与环境。

(3) The meaning of energy and land saving buildings

Energy and land saving building never imply that people should live in a low standard narrow space. On the contrary, it stands for buildings with new science and techniques, buildings fully utilising clean, safe, eternal solar energy and other sustainable energy sources instead of pollutant, unsafe, finite, non-sustainable energy, building environment with beautiful appearance and suitable density, the combinations of aboveground buildings with underground ones, and buildings on the sea and on the land to create wonderful living space and environment for people.

2.3.4 生态可持续危机引发了生态可持续建筑

(1) 对"天、地、人,和"的重新认识

20 世纪 80 年代后期起,由于人口、能源、土地、建房、环境之间的矛盾以及天灾人祸的负面影响,危及全球的生态可持续发展,迫使我们寻求新的对策。其时,建筑科研者(含本书作者)反思发现:节能建筑、太阳房、节能节地建筑,它们的含义及原理均从未明确:建筑应促进含其自身在内的各领域的生态平衡及可持续发展。从那时起,国内外建筑界有识之士乃又开始了"生态与可持续建筑"和"绿色建筑"的研究,并重新想起了我国古代"天、地、人,和"的宏伟宇宙环境观,重新认识到该哲理的历史永恒性,重新领悟到"天、地、人,和""物我相应,谐和共生"的必要性。图 2-11 为该理念示意图。

2.3.4 The crisis of eco-sustainability led to eco-sustainable architecture

(1) Re-understand the "Harmony between Sky and Earth and Man"

Since the late 1980s the contradiction among population, energy, land, housing and environment and the negative impact of natural and man-made calamities have increasingly damaged the eco-sustainable development throughout the world. This forced us to find new countermeasures. From then, the architectural researchers (incl the authors of the book) in reflection discovered that: the meanings and principles in energy saving buildings, solar buildings, and energy-land saving buildings, have never been clearly stated that architecture should promote eco-balance and sustainable development of all fields including itself. Ever since, in architecture field, the researchers and experts at home and abroad have started the research on "Eco-sustainable Architecture" and "Green Architecture". They have realized the existence of the ancient Chinese philosophical views on universe: "Harmony between sky and earth and man", understood the philosophy's historic eternity, re-understood the necessity of "Harmony between sky and earth and man" and "Object and subject suits to each other and harmonically grows together". (Fig. 2-11 showing the idea).

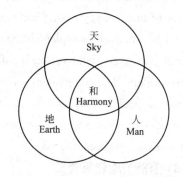

图 2.11 "天、地、人,和"哲理指引我们应努力做到:"物我相应,谐和共生"

Fig. 2-11 "Harmony between Sky and Earth and Man" this philosophy leads us to do the best in "Objects and subjects suit to each other and harmonically grow together"

(2) 生态可持续理念广泛传播

重温"生态"含义:生态就是事物起源、生存、发展的条件和环境。人类生态可持续发展的关键条件和环境是:天界:太阳、月亮、大气圈;地界:含地球本身及其山山水水、矿藏,还有比人类历史长得多的动、植物;人界:人类环境是政治环境、经济环境、文化环境和科技环境组合的多色多元的庞杂社会。积天灾人祸的教训,越来越多的人终于认识到各领域生态可持续性的重要性。许多权威性国际会议相继发号召,搞竞赛,促行动,广泛传播生

2.3 新概念建筑
2.3 New Concept Architecture

态可持续性的理念，例：

1991年，墨尔本"生态设计国际会议"强调各行业都必须考虑生产与生态的关系。

1992年，斯德哥尔摩与赫尔辛基国际建协"生态－建筑国际会议"，重点讨论了生态平衡在建筑中的绝对必要性。

1992年，里约热内卢"全球峰会"，170多国共商社会的可持续发展问题。

1993年，国际建协与联合国教科文组织协组："为可持续的将来进行策划世界大会"及设计竞赛。

1993年，布达佩斯"世界太阳能大会"强调人与自然谐和。

1994年6月5日"世界环境日"提出"世界一家"的号召。

1997年，韩国·济州岛"绿色美景世界大会"号召为人类创造富有生机的绿色美景。

1999，北京，国际建协"21世纪建筑国际大会"主旨就是生态可持续发展。

2008北京奥运会口号"同一个世界，同一个梦"，多好的口号！

回顾以往，最深刻的教训是：人类必须醒悟，再不要做伤害自然最终伤害自己的负效行为。各领域都应为人类社会生态可持续发展作出更好的贡献。

(2) Wide spreading of eco-sustainable idea

Reviewing the meaning of "Ecology": "Ecology" means the conditions and environment of things' origin, existence and development. The key conditions and environment for human eco-sustainable development are: sky: including the sun, the moon and the atmosphere; land: including earth itself and its mountains, rivers, minerals, as well as animals and plants whose history is much longer than that of humanity; human: human society are a colourful and diversified complex society composed of political, economical, cultural and science-technical environment. Under the accumulative lessons of natural and man-made calamities, more and more people finally have realised the importance of the eco-sustainability in all areas. Many authoritative international congresses one after another have issued calls, organized competitions, promoted actions for widely advocating the idea of eco-sustainability. For examples:

In 1991, in Melbourne, "Eco-design International Congress" stressed that all trades and professions must consider the relation between production and ecology.

In 1992, in Stockholm & Helsinki, UIA "Eco-architecture International Congress" focused on discussing the absolute necessity of eco-conservation in architecture.

In 1992, in Rio de Janeiro, "Earth Summit" had more than 170 countries' representatives together discussing the sustainable development of our society.

In 1993, UIA and UNESCO cooperatively sponsored the "Design for Sustainable Future World Congress" and design competition.

In 1993, in Budapest, "Solar Energy World Congress" stressed "Harmony With Nature".

On June 5th 1994, the "Word Environment Day" suggested the call: "One World, One Family".

In 1997, in Korea, "Green Enertopia World Congress" called for all fields to work hard for human green enertopia future.

2 温故知今与创新——中国建筑回顾
2 LOOKING BACK TO THE PAST TO UNDERSTAND TODAY AND CREATE THE NEW——Reviewing the Chinese Architecture

In 1999, in Beijing, UIA "International Congress of the 21st Century Architecture" had its major focus on Eco-sustainable Development.

In 2008, Beijing Olympic Games' calls "One World, One Dream", a very good call!

Reviewing the past, the hardest lesson is that human beings must wake up to stop the negative behaviours harming nature and finally harming ourselves. All fields should contribute much better for the eco-sustainable development of human society.

3 21世纪建筑何往
3 WHERE DOES ARCHITECTURE GO IN THE 21ST CENTURY

3.1 必然与偶然的反思

在回顾建筑的历程中,我们已经概括了这样一条历程:传统建筑→节能建筑、太阳房→节能节地建筑→绿色建筑;生态可持续建筑。这是一条必然的过程吗?……非也!这恰恰说明:我们的观念总是落在"危机"后面。

3.1 Looking Back on Necessity and Contingency

Looking back to the architectural history we have summarized its evolution process as the following: traditional buildings→energy saving buildings and solar buildings→energy and land saving buildings→green architecture and eco-sustainable architecture. Is this an inevitable course? … No! This just explained that our concepts were behind the "crisis".

能源危机暴露了传统建筑浪费能源的缺点,这才激起我们去研究、实验、推广节能建筑。

当土地危机来时,建筑是占地最大户,这又激发我们思考:建筑只是节能还不够,必须同时节能节地,于是在研究领域又出现节能节地建筑。

能源、土地、绿色危机与多种污染已经并仍在阻碍可持续发展时,乃又兴起了绿色建筑和生态可持续建筑的研究并成为热门课题。

拿可持续建筑的要求来衡量,那么上述建筑(仅居住建筑已超400亿 m²)都必须改造。这难道不值得深思吗?

"少一点落后实际的被动回顾","多一点预见(超前)性思考",必将使我们获得先进的科学成果。当这些成果转化为生产力时,比上述被动式发展必将作出更多的贡献。

Energy crisis exposed conventional buildings' shortcomings of wasting energy therefore stimulated us to study, experiment on and implement energy saving buildings.

When land crisis came, buildings are the biggest intruders occupied land. It stimulated us again to consider that it's not enough that buildings can only save energy. It must save energy and land at the same time. Then in research field energy saving and land saving buildings emerged.

Since the crises of energy, land, green and various pollutions became barriers to the globally sustainable development, the researches on green architecture and eco-sustainable architecture have been emerging and becoming the fashionable topics.

If using the sustainable architecture standards to evaluate the conventional buildings (in China there are over 40 billion m² dwellings alone), they all have to be refurbished. Is this not worth pondering?

"To make less passive reviews behind practice", "and have more leading (foreseeing) thoughts" would make us open to advanced science results. When these results have been converted into productivity that certainly would give more contributions than what the above-mentioned passive review can do.

3.2 建筑含义的延伸

建筑的原始含义就是"庇护所",从天然洞穴、人工穴居、古代小屋都早已显示了这一功能。现代,千姿百态的建筑已成为万紫千红世界空间的主体。但其本质含义仍是"庇护",只不过是从古代简陋的庇护进展到了现代科学较完善的庇护而已。

美国大百科全书已将建筑含义延伸:建筑不仅指房屋,还包括工厂、道路、桥梁、河港、码头、海湾、运河、航空港等建筑工程❶。该书只是延伸了地球传统建筑含义,21世纪建筑含义从空间范围必将向太空、海洋、地下延伸,这些建筑的质量必将从传统建筑、节能建筑、太阳房、节能节地建筑的水平向21世纪的生态可持续建筑、绿色建筑的质量水平提高。

3.2 Extending the Meaning of Architecture

The original meaning of architecture is a "shelter". This function has been already demonstrated in the natural caves, artificial cave dwellings and small ancient wooden houses. Today, the buildings in various types and styles have become the main body of the colourful three-dimensional world. But their essential meaning still is a "shelter". The difference is only between a simple and primitive ancient shelter to a better shelter by modern science and technology.

Encyclopedia Americana has broadened the meaning of architecture as: Architecture refers to not only house but also the building projects of factory, road, bridge, port, wharf, bay, canal, airport and etc. ❶ This book only broadened the meaning of the traditional architecture on earth. In the 21st century, architecture is going to broaden its meaning to include buildings in space, sea and the underground space. Their quality is to be developed from the standards of the conventional architecture, energy saving building, solar building, energy and land saving building into that of eco-sustainable architecture and green architecture of the 21st century.

---

❶ 王宏经,何健主编·中国当代建筑论坛《前言》山东大学出版社,1997

❶ (Preface of Contemporary Chinese Architecture Forum. Shandong University Press. 1997. Editors: Wang Hongjing, He Jian).

3.3 21世纪建筑前景

3.3.1 上天——发展太空建筑

一个世纪是100年,预言21世纪太空建筑必将发展,并非科幻梦语。

(1) 登月

据信息,美国曾有计划:2050年移民月球10万人。欧洲航天局也计划40年里送人和机器上月球,为移民作准备。美、法两国工程师正研究在月球就地取材建月球建筑(从地球运水泥到月球,每吨运价5000万美元)。美国由一位美籍华裔年轻工程师制成了月球尘土水泥样品。国际宇航会急发呼吁"月球应成为全人类的共同财富,应大家共管",唯恐美、法先登,将月球占为太空殖民地。2006年6月25日多国订约重申月球应属于全人类的主张。

3.3 Outlook of Architecture in 21st Century

3.3.1 Go into sky-to develop space architecture

One century has 100 years. To forecast space architecture's development in the 21st century is certainly not a science fiction.

(1) Man on the Moon

It was reported that USA is planning to emigrate 0.1 million Americans onto the moon in 2050. The Space Department of Europe also plans to send people and machines onto the moon to prepare for emigration during the coming 40 years. The engineers of both America and France have been studying to build moon buildings with local materials on the Moon (To carry cement from earth onto moon, the cost is US＄50m/t). An American young engineer of Chinese origin has completed a cement sample made of moon dust. The International Space Society has made an urgent appeal: "The moon should be the common property of all humanity and managed by all countries", based on the concern that America and France will firstly occupy the moon as a space colony. On July 25th 2006, many countries have reached an agreement to reaffirm: Moon should belong to all mankind.

移民就意味着在移居地永久性定居,并生儿育女,传宗接代。移民月球就意味着将地球人变为月球人。此举谈何容易!

地球人,正确地说,应是地表人,其生理运行规律与地表生态环境(太阳、大气圈、空气成分、压力、重力、水、绿色世界……)密切相关。仅举一例:图3-1为地球人血液中元素与地球岩石元素对照表。两者的元素类型、含量竟达到如此惊人的同步波形!地球表面约71%是水(面积比),人体65%~70%也是水(重量比)。以上表明地球人与地球的关系是多么密切,"地球是地球人最伟大的母体",此话毫不为过。

Emigration means people moving to a new place where they'll live and bear children from generation to generation. Emigrating into the moon means that the earthman would become the moon man. This is easier said than done!

The earthman, to be precise, should be the man on the earth surface, whose physiological living pattern is closely related to the ground eco-environment (the sun, atmosphere, air component, pressure, gravity, water, green world, and etc.). Fig. 3-1 is a diagram of a comparison of the elements in earthman's blood sample and the ones in the earth rocks sample. The types and contents of elements in both samples are in such surprisingly synchronized waves! On the earth's surface, water has occupied 71% of the total area (area ratio). 65%~70% of the human body on earth is water (weight ratio). The above shows that the relation between the earthman and the earth is so close that it is not an overstating, "the Earth is earthman's greatest mother".

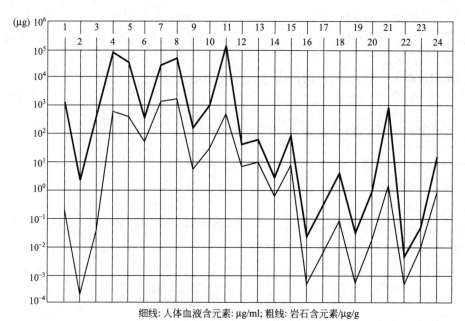

细线: 人体血液含元素: μg/ml; 粗线: 岩石含元素/μg/g
Thin line: Human blood contained elements μg/ml
Thick line: Rock contained elements μg/g

图 3-1 地球人血液元素与地球岩石元素对照

Fig 3-1 The comparison of the elements between what are in the earthman's blood and in the earth rocks.

1—氢 hydrogen; 2—铍 beryllium; 3—氟 fluorine; 4—钠 sodium; 5—镁 magnesium; 6—磷 phosphorus; 7—钾 potassium; 8—钙 calcium; 9—铬 chromium; 10—锰 manganese; 11—铁 iron; 12—铜 copper; 13—锌 zinc; 14—砷 arsenic; 15—镓 gallium; 16—铑 rhodium; 17—镉 cadmium; 18—锡 tin; 19—碲 tellurium; 20—碘 iodine; 21—钡 barium; 22—铼 rhenium; 23—汞 mercury; 24—铝 aluminium

(资料来源: 张宜仁教授提供: 中国环境卫生学百科全书)

(Source: Prof Zhang Yiren supplied Chinese Encyclopedia of Hygienic)

根据"万星同源"学说，今日宇宙所有星球均源于150亿地球年前一次宇宙大爆炸。故月球岩石元素与地球岩石元素可能相同或相似。但是两者之间生态环境却截然不同：月球表面重力仅为地表重力的1/6，无空气、缺水，无绿色植物，白天温度高于130℃，晚上零下180℃。因此，大规模移民月球，难！但是，在月球上建立科研站、旅游建筑，甚至创造生活环境进行先动物后人类生儿育女的试验基地都是可能的。

3.3　21世纪建筑前景
3.3　Outlook of Architecture in 21st Century

According to the theory of "All stars are from one origin", all stars in today's universe originated from one Big Bang happened 15 billion earth years ago. So the rocks' components on the moon and on the earth might be the same or similar. But, the eco-environments between the two are indeed opposite to each other. The moonscape gravity is only 1/6 of that of the earth surface. On the moon there is no air, few water, no green plant. The daytime temperature on the moon is higher than 130℃ and the drops 180℃ below zero at night. Therefore, large-scale emigration onto the moon is difficult! But, it might be possible to build research stations, tourist buildings, or even create living environment for animal test first, man later to bear children on the moon.

对于我国来说，即使能移民100万到月球，对缓解我国人口过多问题，功效实是微乎其微。然而，上月球搞科研如同去地球南、北极科研一样，我们决不放弃。月球无空气，形成不了"风"，其表面尘埃不会飞到月上空间降低能见度。在月球上天文望远镜分辨力可达原"哈勃"太空望远镜10万倍以上。月球是一个理想的太空科研站，同时还有60多种矿物资源。(特别是氦3——一种高效、清洁、安全的核聚变发电燃料，100t氦3便能提供全球一年所需能源总量)。中国嫦娥2号探月卫星已成功发射，为嫦娥3号、4号选定最佳登月点。

For China, even if one million people were emigrated onto the moon, there are very few benefits to alleviate the problem of over crowded population. But, we will never give up to carry out research on the moon the same way as what we are doing in both South Pole and North Pole on the earth. On the moon, there is no air. So there is no wind therefore the moon dust can't fly up into its sky to reduce the visibility. The resolution of the space telescope is 100 thousand times more than that of the original Hubble Space Telescope on the earth. The moon is an ideal place for space research station and has more than 60 types of minerals. (Especially, rich in the He-3, a high efficient, clean, safe nuclear fusion fuel in generating electricity. 100 tons He-3 can supply a whole year's earth energy needs). a Chinese satellite, Chang'e 2, exploring the moon, has been successfully launched to look for the best landing spots for Chang'e 3 and Chang'e 4.

(2) 月球与地球的其他关系
① 维持地球正常运行
月球是在太阳系形成过程自然形成的地球的卫星。所有星球的运行都处于彼此引力与离心力作用的动态平衡中。
设若没有月球，地球将滚翻运行，四季乱序，地球生物难以生存。

(2) Other relationships between the earth and the moon
① Keeping the earth moving normally
The moon is a natural satellite of the earth. It was naturally formed during the forming process of solar system. All stars move in equilibrium by their gravities and centrifugal forces acting together.

If there were no moon, the earth would move in roll and toss, and the changing pattern of the four seasons would be in disorder and all living things would be hard to live.

② 潮汐作用

月亮的引力使地球海水不能与地球自转方向同步运行，而是相对于海底逆向而行，形成潮汐。

利用潮汐效应蓄水发电是一项可再生的可持续能源。

② Tidal effect

The Moon's gravity forces the seawater on the earth not move in the same way as the direction of the earth revolving on its own axis. So the seawater and seabed move in opposite direction from each other, hence causes tides.

Using tidal effect to store seawater to generate electricity is a renewable sustainable way of generating energy.

③ 地震影响

月球对地球的引力对地震也有影响，有关专家一直在研究。当然，地球引力同时也会影响月震。

③ Earthquake effect

The Moon's gravity also impacts on earthquakes. The relevant specialists have been studying the influence. The earth gravity can also influence moonquake too.

④ 保护地球

当有某星体可能撞击地球时，从月球发射武器予以击碎或迫其改变轨道。

④ Protecting the earth

If a star collides into the earth, we can fire from the moon to break it or force it to change its orbit, to protect the earth.

⑤ 有益于地球人的夜生产与精神生活

月亮对地球人的精神生活自古以来就扮演着重要的角色，仅诗仙李白一人就有多篇吟月诗句，例：

《月下独酌》 （引自英汉对照《唐诗三百首新录》）

花间一壶酒，独酌无相亲。

举杯邀明月，对影成三人。

《静夜思》 （引自英汉对照《唐诗三百首新录》）

床前明月光，疑是地上霜。

举头望明月，低头思故乡。

披星戴月行路人，边疆、海岛战士们，月下歌舞儿女们，海洋作业军、民们……多少人在享受这免费天灯惠施的温柔乳白的照明！

⑤ Benefiting earthman's production and cultural life at night

Since ancient times, for earth man's cultural life the moon has played an important role. The famous ancient poet Li Bai alone had written a lot of poems to sing the moon. For examples:

Drink Alone under the Moon (extracted from English-Chinese Poems From Tang Dynasty)

Amid the flower, from a pot of wine l drink alone beneath the bright moonshine. l raise my cup to invite the Moon who blends her light with my shadow and we're three friends.

3.3 21世纪建筑前景
3.3 Outlook of Architecture in 21st Century

Thoughts in the Silent Night (extracted from English-Chinese Poems From Tang Dynasty)

Beside my bed a pool of light, is it hoarfrost on the ground? I lift my eyes and see the moon, I bend my head and think of home.

Passengers under the canopy of the moon and the stars, soldiers guarding at the borders and remote islands, men and women singing and dancing under the moon, army and people working on the ocean…so many people have been enjoying the smooth and milky lighting vouchsafed from the free sky lamp!

⑥ 太空旅游宫、宇宙研究站

⑥ Space tourist palace, universal research station.

(3) 发展其他太空建筑

宇航学、生物学、农学、医学等太空实验研究已有相当基础。我国神舟3号、4号、5号、6号、7号宇宙飞船都曾进行多项有关上述太空实验，如实验的青椒太空种子，在太空条件［高洁净、高真空、高辐射、微重力（三高一低）］处理后，回地面种出的青椒每只平均5～7两（250～350g），比原地面青椒重3倍以上。

预期21世纪各种太空实验室、太空观测站、太空旅游站等太空建筑必将大发展。

(3) Developing other buildings in the space

In space, the experimental studies of aerospace science, ecology, agriculture and medicine have had a good start. Chinese spaceships: Shenzhou NO. 3,4,5,6 and 7 all have carried out many projects of the above experiments. For example: the green pepper's seeds treated under the space conditions (high degree hygiene, high vacuum, high radiation and microgravity-three "high", one "low"), were brought back and planted on ground. The grown pepper weighed at average 250～350g each. This weight was more than three times of the original ground pepper.

We expect, during the 21st century, various space buildings, such as space laboratories, space observation stations, space tourist stations will be greatly developed.

3.3.2 下海——发展海洋建筑

日本某建筑公司曾拟在近日本海域水深100m处，建造一周长100km、直径32km的环形水坝，抽出海水，建造一座底面为700km^{2}的居住城，容纳200万人生活；又有一群工程师和建筑师设计在离东京120km、水深200m处，用1万根柱桩支承一座钢架平台，其上建一座100万人的"海洋城"；中国辽宁拟建1亿亩面积的"海上辽宁"（1亩＝0.0667公顷）；福建拟建1600万亩的"海上林园"；浙江拟建9000万亩的海上绿色工程；山东也有海上发展规划；澳门长期填海造陆近10km^2。其实，水上建筑早已有之：渔船、海轮、军舰无不是水面游动建筑；潜艇则是水中游动建筑；海面油、气作业平台则是海洋固定建筑。三国时代赤壁之战，诸葛亮与周瑜合战曹操，火烧联锁甲船曹军83万，该联锁甲船就是一座庞大的水上游动兵营。

海洋面积占地球表面的71%。人类向海洋要能源、矿资、绿色面积、建筑空间的时代已经来到。生物学、生态学的专家们早已呼吁，发展海洋绿色植物。利用其光合作用吸收CO_2、提供O_2，比陆地植树造林时间短，见效快，面积大，没有旱涝灾害，没有水土流失和森林火灾以及人为毁林等祸害。海洋绿色植物是平衡大气温室效应，防止气候过暖，促进全球生态平衡的最好途径之一。仅此一项海洋绿色工程就需要大量海洋建筑供人居住、科研、生产、经营、管理。此外，还有大量海洋矿源工程，能源工程……。因此，海洋建筑必将大发展。

3.3.2 Go into the sea-to develop marine architecture

A Japanese construction company has designed a city on the sea near Japan for two million people, where the depth of seawater is 100m. They designed to construct a circular dam with perimeter of 100km and diameter of 32km. Then the water is pumped to build a city with footprint of 700 km^2. Another group of engineers and architects have designed an "Ocean City" on the sea, 120km from Tokyo, where water depth reaches 200m. It has a steel framed platform supported by 10,000 piles to prepare an Ocean City for 1 million people. Liaoning Province in China is going to build a "Liaoning on the sea" city with 0.1 billion Mu area (1mu=0.067ha). Fujian, is to build a "Sea Park" of 16 million mu area. Zhejiang is to build a green project of 90 million Mu area on the sea. Shandong also has a plannihg development on the sea. Macao has created land area near 10 km^2 through a long-term work of filling the sea with soil. In fact, buildings on the water have already existed: fishing boats, ships, war ships are all moving buildings on water. Submarines are moving buildings in water. The platforms on the sea producing oil and gas are fastened sea buildings. During Three Kingdoms' Chibi War, Zhu Geliang and Zhou Yu cooperated to fight Cao Cao by burning Cao's force of 0.83 million soldiers on the boats linked together. The linked boats were indeed a very big moving barracks.

The ocean occupies 71％ area of the earth surface. The time is coming for mankind to harvest energy, minerals, green area and building spaces from the sea. The experts of biology and ecology have already called to develop sea green plants and use their photosynthesis to absorb CO_2, supply O_2. Comparing with land afforestation, this has the benefits of shortening time, quick gains, vast area, no drought or water logging disasters, no soil erosion, no bush fires and no deforestation by man, and etc. Relying on the sea green plants are one of the best ways to balance atmospherical greenhouse effect, eliminate climate over warming, and increase global eco-balance. Only for this green project on the sea, there are a lot of demand on buildings for living, researching, producing, managing and administration. Besides, there are much more mining projects, energy project and so on. It is certain that the marine architecture would be greatly developed.

3.3.3 陆地建筑仍是主要方向

全球人口已超 70 亿（中国已超 14 亿），可以说都住在陆地上。中国 21 世纪陆地建筑必须抓好下列四点：

1) 大力开发、利用地下空间，特别是浅层地下空间；
2) 按生态可持续建筑要求管好城乡每年的新建筑（仅新居住建筑每年就有约 20 亿 m^2）。
3) 按生态可持续建筑要求改造好现有建筑（仅住宅就有约 400 亿 m^2）——一项十分艰

3.3 21世纪建筑前景
3.3 Outlook of Architecture in 21st Century

巨的任务!

4) 大力发展多层可种植温室,更多增加农艺和园艺产品,使人民物质生活和精神生活的物质资源更加富足。

太空建筑、海洋建筑以及陆地建筑(含地下建筑)都必须符合生态可持续建筑的理念与原理。

3.3.3 Land architecture-still the main direction

The global population has been over 7.0 billion (China has been over 1.4 billion). Almost all are living on land. The 21st century Chinese land architecture must pay attention to the following four points:

1) Greatly developing and using underground spaces, especially shallow underground spaces;

2) Well managing the new annual constructed buildings in urban and rural areas according to eco-sustainable architecture (ESA) standards because just the new dwellings developed annually are about two billion m^2;

3) Well refurbishing the existing buildings according to ESA demands. The existing dwellings alone have reached about 40 billion m^2. It's a very hard job!

4) Greatly developing multi-story plantable greenhouses to increase agricultural & horticultural products to make the material resources richer for man's material life and cultural life.

Space architecture, marine architecture and land architecture (incl underground ones) all must be developed in accordance with the idea and principles of ESA.

4 何谓生态可持续建筑
4 WHAT DOES ECO-SUSTAINABLE ARCHITECTURE MEAN

4.1 可持续性并不是新概念

万事万物自始就存在可持续发展问题，例："不孝有三，无后为大"是中国自古传宗接代的"民族持续发展经"；

新陈代谢是生物持续发展的生理规律；

大气臭氧层是保护地球及其生物生存并持续发展的必要条件之一；

军事大战，商业竞争更是反映了双方持续发展的矛盾（当然有正义与非正义、公平与黑心阴谋之分）。

保证可持续发展的资源必然含物质和精神两方面，相辅相成，缺一不可。国际建协（UIA）与联合国教科文组织（UNESCO）的一份文件对"资源"的解释是："就一个社会来说，其资源可以是物质的，如化石燃料、土层；天然的废物吸收系统，如沼泽地带、大气。可以是社会性的理念，如教育水平和公平竞争，以及开放的、光明磊落的意识"。

我们追求的生态可持续建筑又是如何呢？

4.1 Sustainability is not a New Concept

Sustainable development exists in all matters since their beginning. For examples: "There are three impieties for man. No offspring is the most serious one" is the traditional concept for "sustaining nations from generation to generation" in ancient China.

Metabolism is the physiological law of living things for sustainable development;

The atmospheric ozonosphere is one of the essential conditions for protecting earth and the existence and sustainable development of its living things;

Military wars and business competitions particularly reflect the sustained contradictions between the two sides (of course there are difference between justice and injustice, equity and evil mind).

The resources ensuring sustainable development must contain both material and spiritual part. Both are complementary to each other. Neither can work without the other. A published document by UIA and UNISCO has defined the "resources" as: "To a society, its resources can be materials such as fossil fuel, soil, natural absorbing wastes systems such as marshlands, air; it also can be social ideas such as education level, equitable competition and open and honest ideas".

What is the eco-sustainable architecture we are seeking?

4.2 非可持续建筑

在此,请允许用反证法说明下列建筑都不是可持续建筑:

(1) 使用污染性,不安全能源如煤、气、油、柴、火电等。

(2) 使用有害材料如致癌、过敏、窒息、辐射污染等。惨例:1996 年,一留美归国博士与 9 位同窗好友在济南一餐馆欢庆回国。不幸餐馆失火,10 位英才全被装饰材料有害气体窒息致死)。

(3) 现有建筑反射大量太阳能,排走大量雨水,极大的浪费!作者估算:设全国 2.5 亿户,每户日照面积 50 平方米,按平均水平,每年每平方米日照面可得太阳能 544 万 kJ,取 50% 反射率,按最低光电转换率 11% 计,那么,每年就有 10389 亿度电(每户每年 4000 多度电)被浪费!

(4) 装有非可持续设备的建筑也不是可持续建筑。以夏季使用机电空调为例:这种设备实际上就是热、湿搬运器,将室内热量通过冷媒搬到室外,将室内空气内水蒸气变成凝结水搬到室外。又如排烟机,乃是将炊事油烟搬到室外污染大气的设备。其缺点有五:

① 热污染并加剧热岛负效应。机电空调系统在将室内热量搬到室外的同时,还会产生机器运转的摩擦热,故总热量非但未减少反而有所增多,造成热污染并加剧热岛负效应。粗估全国夏季空调抛弃的凝结水可能超过几十亿吨!全球总量更惊人。

② 破坏臭氧层。前(1.3.1(2))已述及,机电空调等排放的氯氟烃在太阳的紫外线作用下会释放出氯(Cl)原子。一个氯原子能分裂约 10 万臭氧分子(O_3)为氧分子(O_2)和氧原子(O),从而破坏臭氧层。

③ 造成夏季用电高峰。

④ 噪声污染。

⑤ 室内新鲜空气减少。

4.2 Non-sustainable Architecture

Here, let's use examples of non-sustainable buildings to disproof the idea:

(1) Using pollutant, unsafe energy sources: coal, gas, oil, firewood and thermal power (coal-burning) electricity, and etc.

(2) Using materials containing harmful substances such as carcinogen, allergen, suffocation, radiant pollution, and etc. (A miserable case: In 1996, a PH.D has invited nine schoolmates to dine together at a restaurant in Jinan for celebrating his returning home from US. Unfortunately, the building had a fire. The ten intelligent young men all were suffocated to death by the harmful gases from the decorative materials).

(3) Existing buildings reflects most solar energy, drains off large quantity of rainwa-

ter. Their wastage is serious! The author estimates that if assuming each Chinese home has $50m^2$ sunshine area and average each m^2 sunshine area can receive annual solar energy at 5.44 million kJ, assuming reflectance 50% and the lowest light-electricity conversion rate 11%, then all 0.25 billion Chinese homes would have wasted 1,038.9 billion kWh electrical power every year (each home wasted more than 4,000kWh)!

(4) Buildings installed with non-sustainable equipment are also non-sustainable buildings. For example: in summer we use mechanical air conditioning system. In fact, this type of machine is a heat and humidity carrier carrying indoor heat outside by cooling media while condensing the water vapor within indoor air into water to be moved outside. Another example is the exhaust blower, which is a pollutant equpment by carryihg cooking lampblack to outdoor. There are five shortcomings:

① Heat pollution and aggravating thermal island negative effect. Mechanical air conditioning moves indoor heat outside while the machine produces friction heat in operation so the total heat is not decreased but conversely increased and causes heat pollution and aggravates thermal island negative effect. Here is a rough estimate: the abandoned condensed water by the summer air conditioning might be over tens of billion tons in China! Throughout the world the amount would be even more astonishing!

② Damaging ozonosphere. The above (1.3.1(2)) has mentioned that the mechanical air conditioning discharges the hydrochlorinefluor which under the solar ultraviolet rays' action, can release chlorine atoms (Cl). One chlorine atom can break about 0.1 million ozone molecules (O_3) into oxygen molecules (O_2) and oxygen atoms (O). As the result of this process the ozonosphere is damaged.

③ Causing summer electricity peak load.

④ Noise pollution.

⑤ Reducing indoor fresh air.

4.3 可持续建筑含义

建筑及其环境若能做到有利于：综合用能，多能转换，三向发展，增效资源，自然空调，立体绿化，生态平衡，智能运行，弘扬文脉，素质培养，持续发展，美感、卫生、安全。(12条，50汉字)这种建筑可称生态可持续建筑，或简称可持续建筑或绿色建筑。上述12条中，每一条都含有一系列的科研工作，现扼要论述如下：

(1) 综合用能

1) 大力开发利用可持续能源，尽快以可持续能源完全取代现用非可持续能源。

可持续能源基本特征：洁净、安全、永久性。例：太阳能、氢能、风能、水动力能（水重力能、潮汐能、波浪能……）；生物质能例如：沼气、绿色植物能，地热能、核聚变能（常温核聚变发电、超导技术输电是人类能源革命的梦想）。

非可持续能源基本特征：污染、不安全、有限性。

例：现用煤、气、油、柴、火电。

2) 回收利用弃热
① 室内弃热：85%以上的炊事弃热以及空调等设施的散热。
② 人体散热：特别是学校、超级市场、影剧院、体育馆等存在大量人体散热。
③ 生产弃热：钢铁、化工、纺织、电力、材料等企业，正惊人地浪费大量弃热。
④ 人体动能
日本：推旋转门余能发电。
伦敦及温哥华冬奥会：跳舞动能发电。
台湾：减肥脚踏车发电。
普通自行车行进中发电早已有之。
2010上海世博会，许多展馆都展示了大量综合用能的出色范例。

4.3 The Meaning of Sustainable Architecture

If the architecture and its environment can bring benefits to: comprehensive use of energy; conversion of multiple form of energy, three-dimensional development; increasing efficiency of resources; natural air conditioning; stereoscopic landscaping; eco-balance; intelligent operation; developing cultural context; fostering quality; sustainable development; beauty, health and safety. (12 items, 50 Chinese characters), we may call it eco-sustainable architecture or simply call it sustainable architecture or green architecture. In the above 12 items, each contains a series of researches. We can summarize them as follows:

(1) Comprehensive use of energy

1) Greatly developing and using sustainable energy, to replace non-sustainable energy as quickly as possible.

The basic characters of sustainable energy are clean, safe and long lasting, such as: solar energy; hydrogen energy; wind energy; hydropower energy (hydro-gravity energy, tidal energy, wave energy…); biomass energy such as methane (biogas; marsh gas; fire-damp), green plant energy; geothermal energy; and nuclear fusion energy (electricity generated by nuclear fusion at room temperature and transmitted by superconductor. That's a human dream in energy revolution).

Non-sustainable energy's basic characters are: pollutant, unsafe, finite. For examples: coal, gas, oil, firewood and coal-burning power.

2) Reusing abandoned heat

① Indoor abandoned heat: more than 85% of cooking heat, and the diffused heat from air conditioning and other equipment.

② Diffused heat from human bodies: especially in schools, supermarkets, cinemas & theatres and gymnasiums, and etc. There exists large amount of diffused heat from human bodies.

③ Abandoned heat from production: industries of steel, chemicals, textile, electrical power and materials, and etc. are wasting astonishing amount of abandoned heat.

④ Human body kinetic energy:

Japan: Using the remaining energy of pushing revolving (revolution) door to generate electricity.

London and Vancouver Winter Olympic Games: Using remaining energy from dance movements to generate electricity.

Taiwan: Using exercise bicycle to generate electricity.

It is also known to use ordinary bicycles in running mode to generate electricity.

In the 2010 SHANGHAI EXPO, many exhibition pavilions have demonstrated outstanding examples which used energy comprehensively.

(2) 多能转换

1) 多功能构件

例：太阳能加热器和太阳能电池板与屋面、外墙面相结合将太阳能转换成热水、电力。用屋顶花园、墙面绿化、阳台绿化、窗台绿化和散水绿化将太阳能转换成绿化能。

2) 一房多用

例：一次作者在美国明尼苏达大学开国际会议。大会主席报告完了分组讨论，不一会，活动隔声隔断，从顶棚降到地面，将大会议厅分隔成若干小会议室，分组讨论在原场进行，省时且高效使用了建筑空间。这是一房多用的好例子。

传统建筑发展表明：随着人们物质与精神生活越来越多样化，建筑类型、房间类型也越来越多样化，形成一种思维模式：物质与精神生活的多样化必然引起建筑及房间的多型化。实践证明：结果是建筑类型增多，建筑总量加大，而建筑使用效率却很低。例如学校、影院、剧院、体育馆、办公楼，每天只使用8～10小时，若该建筑使用寿命为100年，其间即有50多年空而未用！

对策：一房多用。措施：

① 大空间轻构件灵活分隔；

② 建筑空间多功能：例：学校教室、图书馆、实验室除服务本校外，向社会开放，为普及教育服务；体育馆、影院、剧院均可在空余时间对社会开放，服务科、教、文、艺。

一房多用做得好，有利于节能、减排、节地、节材、节工、节时、节财与环保，并减少市政设施。

(2) Multi-form of energy conversion

1) Multi-function components

There are examples of solar heaters and solar cell panels combined with roofing and external walls to convert solar energy into hot water and electricity. Landscaped roof garden, wall, balcony, windowsill and apron convert solar energy into green energy.

2) Multiple use of one building

Once the author was attending an international conference in Minnesota University, USA. After the chairman's report, attendants were divided into groups to discuss. In a few minutes, the movable sound insulation partitions from ceiling falling to floor have divided the big hall into a number of smaller meeting rooms. The groups' discussions could be held at the original hall. Time was saved and the building space was efficiently used.

4.3 可持续建筑含义
4.3 The Meaning of Sustainable Architecture

This is a good example of multi-use of one room.

Conventional architecture development shows that: as people's material life and cultural life are getting more and more complicated, building types and room types are also getting more and more multiformed. Following a thinking pattern people believe that more complicated material and cultural lives would certainly lead to more multiformed buildings and rooms. Practice has proved that: as a result of this typical thinking, the type and number of buildings have increased, but the building occupation rate has become very low. For examples: schools, cinemas, theatres, gymnasiums and office buildings are only occupied 8 to 10 hours each day. If the building's life span were 100 years, it would be empty for more than 50 years!

The Countermeasure is: multiple use of a building. The strategies are:

① Dividing large space with light movable partitions;

② Creating multi-functional building space. For examples: Making school classrooms, libraries, laboratories, available not only for their own institutions, but also opened to society for public education. Gymnasiums, cinemas, theatres all can be opened to the public after the normal business hours to be available for science, education, culture and art.

If well organized, a building with multiple function can benefit energy saving and land saving, reduce harmful discharges, save material, time, finance, protect environment and reduce public infrastructure and facilities.

(3) 三向发展

除以发展陆地建筑为主外,还应发展太空建筑、海洋建筑和地下建筑。21世纪初、中期,中国建筑应地上、地下相结合,陆地、海洋相结合,尤应注意开发利用浅层地下空间,大力发展各种山地、坡地建筑。

(3) Three-dimensional development

Besides developing land architecture generally, China should also develop space architecture, marine architecture and subterranean architecture. From the beginning to the mid of 21st century, Chinese architecture should be developed in combination of aboveground and underground buildings, land and marine buildings. Particular focus should be placed on the development of shallow underground spaces and various hilly areas and sloping sites.

(4) 增效资源

生态可持续建筑必须有利于增效资源,如:节能、节地、节材、节工、节时、节钱、减排、平衡大气温室效应、消除气候过度变暖。并有利于增多绿色面积及其他环保。

关键点是:使建筑体形系数 F/V、建筑容积系数 F/V_{net}、建筑规划的周长与面积比 P/A 以及建筑布置(BL)达到优化,并提高围护结构冬、夏绝热及透明构件的效率(详见6.增效资源)。(F:建筑外表面积(m^2);V:建筑外形体积(m^3);V_{net}:建筑内净空(m^3);P:建筑规划的周长(m);A:建筑规划的面积(m^2))。

(4) Increasing efficiency of resources

Eco-sustainable architecture should increase efficiency of resources. There are total 10 benefits. They are energy saving (ES), land saving (LS), material saving (MS), work saving (WS), time saving (TS), finances saving (FS), reducing discharge (RD), balancing AGE (BAGE), eliminating COW (ECOW) and increasing green area and other environmental protection (EP).

The key points are: To optimize Building Shape Coefficient (BSC) = F/V, Building Volume Coefficient (BVC) = F/Vnet, the ratio of building planning periphery over area (P/A) and building layout (BL); To increase envelopes' insulation in winter and summer and the efficiency of transparent elements. (F: the external surface area of the building (m^2); V: building external volume (m^3); Vnet: net internal space in the building (m^3); P: building planning perimeter (m); A: building planning area (m^2)).

(5) 自然空调

① 保护、改善自然生态环境，充分发挥绿色植物与江、河、湖、海、沼泽地大环境的自然空调作用。

② 对于建筑，注意利用建筑材料(如水、土)的热、湿惯性构建自然空调系统(实验例见7.3.4自然空调系统)

(5) Natural air conditioning

① Protecting and improving natural eco-environments. Fully utilizing the natural air conditioning effects of the macro-environments including green plants, rivers, lakes, seas and marshlands.

② For buildings, focus on using building materials' (such as water and soil) thermal and humidity inertia properties to build a natural air conditioning system (see 7.3.4, an experimental Natural air conditioning system).

(6) 立体绿化

立体绿化就是使建筑散水、墙面、窗台、挑檐、遮阳板、阳台、屋面以及内、外空间能绿化的尽量绿化，使绿化的建筑群成为环境绿化中立体花园的主体，使建房后绿色面积少损失，不损失，甚至增多(详见8建筑绿化)。

(6) Stereoscopic landscaping

Stereoscopic landscaping means to maximize landscaping building's apron, walls, windowsills, overhangs, shading boards, balconies, roofing, indoor and outdoor spaces as possible as we can so as to make the landscaped building groups become the main bodies of the stereoscopic gardens in the environment. This is to reduce green area loss, have no lose or even increase (Detail in 8 Greening of Buildings).

(7) 生态平衡

前已述及，我国古代的大环境观"天、地、人，和"的理念实是生态平衡的哲学和科学含义相结合的最高准则。

(7) Eco-balance

We have mentioned previously that Chinese ancient view about the macro-environment as "Harmony between sky, earth and man" is the highest norm of the meaning combining philosophy and science with the eco-balance.

生态危害无国界。例如：河流污染、大气污染、臭氧层被破坏、绿色世界被破坏，从

4.3 可持续建筑含义
4.3 The Meaning of Sustainable Architecture

而加剧大气温室效应等带来的大量危害。它们所影响的是群体生态系统直至全球的可持续发展。

Eco-damage has no national limit. Polluted rivers, polluted air, damaged ozonosphere, and damaged green world aggravating atmospheric greenhouse effect have brought us a lot of damages. They are affecting vast eco-systems and the global sustainable development.

就最高标准来说,生态平衡确是全社会、全球的系统工程。

From the highest standard, eco-balance is a project of system from the whole society and the whole world.

就一个地区或一幢建筑而言,其关键是:做好输入、输出物质流的良性循环和能量形式的良性转换。例如:陕西高陵县建过一幢两层的鸡—猪楼,二楼养鸡,一楼养猪,鸡粪处理后给猪吃,猪粪作沼气原料。沼气能用于炊事、照明,沼气渣养殖(种蘑菇……)、肥田。我们建议在该楼向阳面贴建一可种植的毗连温室,以便使植物光合作用吸 CO_2、吐 O_2 与动物吸 O_2、吐 CO_2 构成良性循环。

Concerning a region or a building, the key point is to ensure the mass flow of input and output in good circulation and the energy forms in good conversion. An example in Gaoling County, Shaanxi Province, a two-story chicken-pig building was built with second floor for raising hens and first floor for pigs. Treated hens' droppings are used as pig's food and pigs' feces are deposit as methane materials. Methane energy is for cooking and lighting and methane dregs are used for cultivating mushrooms and fertilizing the soil. We have suggested to build an attached plantable greenhouse at the side facing the sun so as to achieve a good circulation between plants' photosynthesis, absorbing CO_2 releasing O_2 and animals' breathing in O_2 and breathing out CO_2.

不同使用性质的建筑做到物质流的良性循环及能量形式的良性转换还有大量的研究工作要做。

There are further researches to be done to make buildings with different uses to have mass flow in good circulation and energy forms in good conversion.

(8) 智能运行

就是利用电脑对建筑的功能,智能地操作,例如:

① 日常管理;

② 突发事件:火灾、抢、盗、地震等警报与报警;

③ 呼急救,网络购物……;

④ 机器人进行设定的工作(处理信息,接收主人远距离指令……)。

(8) Intelligent operation

Intelligent operation is to use computer to intelligently operate the building's functions. For examples:

① Daily management;

② Alarming and Emergency reporting of: fire, robbing, breaking in and earthquakes;

③ Emergency calls; on-line shopping…

4 何谓生态可持续建筑
4 WHAT DOES ECO-SUSTAINABLE ARCHITECTURE MEAN

④ Working robots by pre-set instructions (data processing, receiving the master's remote orders…)

(9) 弘扬文脉

生态可持续建筑在利用可持续能源、建筑绿化、自然空调、生态平衡等方面,任何民族、任何地区,原理都是同一的。但在建筑空间的组成与形式以及内外装修、环境设计等方面不能搞同一化,而要个性化。使其充分体现各民族、各地区的优秀民风、民俗和历史相沿的文脉。56个不同民族文脉的建筑已经并仍在体现伟大中华建筑的辉煌多彩。

(9) Developing cultural context

The principles of eco-sustainable architecture in using sustainable energy, greening of buildings, natural air conditioning, eco-balance, etc. are the same for any nation and at any place. But the components and forms of building spaces, inner and outer decoration and environmental design factors should not be the same. Individualism can fully express the excellent local styles and folk culture and the historically continuing context of each nation and each place. Fifty-six nations' buildings at different national context have been demonstrating the splendidness and colorfulness of Chinese architecture.

(10) 素质培养

建筑及其环境是人一生接触最多的有形体。在任何地方,首先输入人脑的信息就是当地的建筑及其环境:当你游览杭州西湖、苏州园林、云南丽江、北京颐和园时定会激起对其建筑及环境美的享受,从而唤起爱美的情操;当参观科技馆时,定会感到科技造福人类的强大威力,激起热爱科学的向往。相反,在城市的街道、车站、机场、港口等建筑墙面常被大量广告占领,起了不少负面影响:图4-1某儿童电影院多年的怪兽门面,真不知给孩子们及过往行人什么样的恐怖感!图4-2改换后就好多了;图4-3为匈牙利布达佩斯英雄广场建筑群也被广告污染,景色大减。

(10) Fostering quality

Buildings and their environments are the most tangible matter people experienced during their lives. No matter where you are, the first information coming into your mind is the local buildings and their environments. When you visit Western Lake in Hangzhou, Suzhou gardens, Yunnan Lijiang River and the Summer Palace in Beijing, you would certainly be inspired and enjoy the beauty of the buildings and environments with aroused sentiment on beauty. When you visit science-technology exhibition hall, you would certainly feel that science & technology are the great power bringing benefit to mankind, arouse your deep love for science and technology. Oppositely, on city streets, station squares, airports and ports, etc. advertising posters occupy the building walls can cause a lot of negative influences: Fig 4-1, a monster poster has been placed at the gate of a children's cinema for many years. We don't know what a terrible feeling it's going to give to children and the passing by pedestrians! Fig 4-2 shows an improvement; Fig 4-3 shows that at Budapest Heroes Square the buildings were polluted by posters. The landscape was seriously lost.

4.3 可持续建筑含义
4.3 The Meaning of Sustainable Architecture

图 4-1 广告牌污染建筑

Fig. 4-1 Building polluted by the billboards

图 4-2 上述建筑出入口改进后情况

Fig. 4-2 The building gate has been improved

图 4-3 布达佩斯英雄广场建筑群受广告牌污染状

Fig. 4-3 At Budapest Hero Square the buildings polluted by billboards

图 4-4 西安钟楼广场建筑群受广告污染状（现已改进）

Fig. 4-4 Around Xi'an Bell Tower Square, the buildings polluted by posters (Now. hare been improve)

设想：大、中、小学校园，都有优美的绿化环境，学生步向课堂时，能听到来自花丛的轻雅音乐，能从大视屏得知最新政经、科技、文体、本校新闻，能从林荫道旁及室内走廊两侧见到有益于人民的伟大历史人物的塑像及其简介。这不正在潜移默化培养我们的素质吗！

We may imagine: if the campus of universities and school yards of middle and primary schools all have fine landscaped environments, students going to classes can hear graceful light music from gardens, receive the latest political and economical news, science and technology, culture and sports, and their own school news from large video screens, they walk pass statues of great historic figures, that represent those changed our lives, and their brief introductions at both sides of avenues and indoor walkways. Is this environmental quality not imperceptibly influencing our personalities!

建筑及其内外环境有着不可取代的立竿见影和潜移默化的培养素质的功能。

Nothing can replace imperceptible influence that the buildings and their environment had on culturing people's quality. The impact is instant.

(11) 持续发展

建筑及其环境必须有利于本身及其他领域的可持续发展。

4 何谓生态可持续建筑
4 WHAT DOES ECO-SUSTAINABLE ARCHITECTURE MEAN

生态可持续建筑与其他领域的可持续发展乃相辅相成谐和共生的关系。

(11) Sustainable development

Buiding and its environment has to benefit the sustainable development for itself and for others.

The relationship between the sustainable developments of eco-sustainable architecture and other fields are complementary to each other and harmonically growing together.

(12) 美感、卫生、安全

这里指的美感不单纯是视觉感受,而是指人的眼、耳、鼻、舌、身、肤、脑全感官的美感。参观画展或欣赏音乐,若该环境使你感到口干舌燥,你还能安心看画或听音乐吗?感官之间有连锁互感性,环境设计应注意这一心理生理特点。

(12) Aesthetic (beauty) feeling, health and safety

Here, a sense of beauty means not only for visual feeling but also feelings of all senses: eyes, ears, nose, tongue, body, skin and mind. When you visit an art exhibition or appreciate music, if the environment makes your mouth and tongue dry, can you be comfortable to watch the paintings or enjoy the music? All senses link and respond to each other. In environment design we have to pay attention to this psycho-physiological character.

卫生、安全不仅包含传统的"庇护"内容,如防雨、雪、风、尘、雹、冬失热、夏进热,防火、抗震等,此外还应防热、声、光、气污染,防辐射(放射性辐射、电磁辐射……)、防盗贼以及防视听信息污染等。历史相沿的重型挑战:我们迄今仍未做到使建筑在巨大天灾、人祸面前保证安全的能力!

The definition of health and safety covers not only the function of traditional "shelter", i. e. protections from rain, snow, wind, dust, hail (hailstone), winter heat loss and summer heat gain, fireproofing, shock proofing, and etc. but also protections from the pollutions of heat, sound, light, gas and radiations (radiant radiation, electromagnetic radiation…), and protections from robbers, and pollutions of unwanted audio and visual information. It has always been such a big challenge during human society development that our buildings still have not had the ability of ensuring human safety under serious natural and man-made calamities!

生态可持续建筑的含义是发展的,随着人们物质与精神生活不断丰富和科技水平的不断提高而发展。

Eco-sustainable architecture's meaning is constantly evolving because of our ever-changing richer physical and cultural life and improving science and technology development.

考虑学时有限与当前实用所需,下面将分别论述太阳能、沼气在建筑中的应用,增效资源,掩土建筑,建筑绿化,优化组合,生态可持续建筑的经济效益,共六章。

Considering limited teaching hours and the needs of today's practical use, we are going to discuss the following in six chapters: The Application of Solar Energy and Methane in Buildings; Increasing Resources Efficiency; Earth Sheltered Buildings; Greening of Buildings; Optimal Combination and the Economical Effects of Sustainable Architecture.

5 太阳能与沼气在建筑中的应用
5 THE APPLICATION OF SOLAR ENERGY AND METHANE IN BUILDINGS

5.1 太阳

5.1.1 太阳基本情况

主要成分：氢［H］78%和氦［He］20%；直径：139万km，是地球直径的109倍；质量：2.2×10^{27} t，是地球质量的33万倍，占太阳系总质量的99.89%。能源：核聚变，每秒消耗400万t质量氢。按爱因斯坦质能互换方程转化为能量，$E = mC^2$（m——质量kg；C——光速，30万km/s），可产生36×10^{22} kW功率的辐射能，以电磁波形式向太空传播，到达地球大气层上界的总功率只及太阳发射的总功率的22亿分之一，约16×10^{13} kW，通过大气到达地面的总功率约8.5×10^{13} kW。这就是地球上一切生物起源、生存、发展的最原始的能源。

5.1 The Sun

5.1.1 About the sun

The sun is mainly composed of 78% of hydrogen [H] and 20% of helium [He]. Its diameter is 1.39 million km, 109 times of the earth. Its total mass is 2.2×10^{27} t, 0.33 million times of the earth and 99.89% of the total mass of solar system. In every second the sun consumes 4 million t mass of hydrogen, to convert into energy in nuclear fusion. According to the Einstein equation of mass-energy exchange: $E = mC^2$ (m: mass, kg; C: light velocity, 300000km/s), it can generate 36×10^{22} kW radiative power, which transmits to the space in form of electromagnetic waves. The power reaching the outside of the earth atmosphere is about 16×10^{13} kW, only one 2.2 billionth of total solar emissive power. Through the atmosphere the power reaching the ground is about 8.5×10^{13} kW. This is the original energy source for the origin, existence and development of all earth beings.

5.1.2 应用太阳能的10个"最"

1) 适应性最强。技术高低都可取用。万物起源、生存、发展早已受惠于太阳能了。

5 太阳能与沼气在建筑中的应用
5 THE APPLICATION OF SOLAR ENERGY AND METHANE IN BUILDINGS

2）覆盖面最广，全国范围都可取用。
3）蕴藏能量最多。
4）最清洁。
5）污染性最少。
6）卫生性最好。
7）多能性最优：杀菌，提供热、光、色。
8）光合作用最宜。没有光合作用就没有地球生物，没有绿色世界。
9）安全性最佳。
10）可用期最长，还可用50亿年。图5-1 示意太阳还年轻。

5.1.2 The ten "most" of using solar energy

1) The most adaptable energy. It can be used through high technology or low technology. All beings' origin, existence and development have already been benefited from the solar energy.

2) The widest coverage. It can be used throughout China everywhere.

3) The most stored energy.

4) The cleanest energy.

5) Minimum pollution.

6) The most hygienical energy.

7) Optimal functions: killing germs. Supplying heat, light and color.

8) Most suitable for photosynthesis. No Photosynthesis, no earth living beings and no green world.

9) The safest energy.

10) The longest application period. Solar energy can still last 5.0 billion years. Fig. 5-1 shows that the sun is still very young.

图 5-1 年轻的太阳
Fig. 5-1 the young sun

5.1.3 中国太阳能资源

到达地球的太阳能约 8.5×10^{13} kW，中国获得约每年 10^{16} kWh，相当于 1.2 万亿吨标准煤的发热量。标准煤燃烧值为 7000kcal/kg。1kcal＝4.187kJ。

我国太阳能分布见表 5-1

5.1.3 The solar energy resources in China

The amount of solar energy reaching earth is about 8.5×10^{13} kW. China receives about 10^{16} kWh per annum, which equals to total generated heat from 1200 billion t standard coal. The combustion value of standard coal is 7000kcal/kg, 1kcal＝4.187kJ.

Solar energy's distribution in China is shown as table 5-1.

5.2 建筑中应用太阳能
5.2 Application of Solar Energy in Buildings

中国太阳能分布　　　　　　　　　　表 5-1
Distribution of Solar Energy in China　　Table 5-1

等级 Rating	日照时数/年 Sunshine hours per annum (annual)	年辐射总量 Total annual radiation energy $10^4 kcal/m^2$	地区 Locations	世界相当区 comparable locations in the world	太阳能水平 Solar energy Level
1	3200～3300	160～200	宁夏北、甘肃北、新疆东南、青海西、西藏西 North Ningxia, North Gansu, Southeast Xinjiang, West Qinghai, West Tibet	印度、巴基斯坦 India, Pakistan	最富 The Richest
2	3000～3200	140～200	河北西北、山西北、内蒙古及宁夏南、甘肃中、青海中、西藏南、新疆南 Northwest Hebei, North Shanxi, South Inner Mongol & South Ningxia, Middle of Gansu, Middle of Qinghai, South Tibet, South Xinjiang	雅加达 Jakarta	富 Rich
3	2200～3000	120～140	山东、河南、河北东南、山西南、新疆北、吉林、辽宁、云南、陕西北、甘肃东南、广东南 Shandong, Henan, Southeast Hebei, South Shanxi, North Xinjiang, Jilin, Liaoning, Yunnan, North Shaanxi, Southeast Gansu, South Guangdong	华盛顿 Washington	中等 Middle
4	1400～2200	100～120	湖南、广西、江西、浙江、湖北、福建北、广东北、陕西南、黑龙江 Hunan, Guangxi, Jiangxi, Zhejiang, Hubei, North Fujian, North Guangdong, South Shaanxi, Heilongjiang	米兰 Milan	较低 Rather low
5	1000～1400	80～100	四川、贵州 Sichuan, Guizhou	巴黎、莫斯科 Paris, Moscow	低 Low

从表 5-1 可看出，我国太阳能分布情况，除个别地区外是北高南低，西高东低：

Table 5-1 indicates that the solar energy distribution in China is richer in north than in south and richer in west than in east except in a few special locations.

北 North：$(120\sim160)\times10^4 kcal/m^2 \cdot a$；

南 South：$(80\sim120)\times10^4 kcal/m^2 \cdot a$；

西 West：$(140\sim200)\times10^4 kcal/m^2 \cdot a$；

东 East：$(80\sim140)\times10^4 kcal/m^2 \cdot a$；

5.2 建筑中应用太阳能

5.2.1 太阳能热水与炊事

太阳能热水器的应用在中国就数量而言已居世界前列。但就全国 25000 万户普遍应用相距尚远。太阳能炊事也在日益发展。

(1) 金属管太阳能热水器

图 5-2 为一金属管太阳能热水器例，由集热器、贮水箱和连接管道组成。集热器内有容水金属管，管外涂黑色选择性涂层（一般黑色吸热率高，同时发射散热也高。黑色选择性涂层如镍黑涂层吸热率高于 90%，而发射率只约有 10%），可大大提高吸热

效率。

5.2 Application of Solar Energy in Buildings

5.2.1 Solar hot water and cooking

The amount of solar heaters used in China has been on top of the list in the world. But it is still far from widespread application of solar heaters in the 250 million homes in China. Solar cooking is also being increasingly developed.

(1) Metal tube solar heater

Fig. 5-2 is an example of solar heater that is composed of a solar collector, a water tank and connective pipes. In the collector there are metal tubes containing water. The tubes' outer surfaces are painted black by selective black paint (the general black has high heat absorption and high heat emissivity. Selective black paint such as the nickel black paint has heat absorptivity higher than 90%, yet emissivity is about 10%). It greatly increases heat absorptivity.

图 5-2 金属管太阳能热水器

Fig. 5-2 A metal pipe solar heater

集热器顶面为透明盖板（常用玻璃）造成温室效应，其余周边均为高绝热构造。本例为热压循环式太阳能热水器，贮水箱底部有通向用户的管道，送热水、补充冷水。

The top surface of the collector is covered with a transparent sheet (usually glass) so as to generate greenhouse effect. The rest sides of the collector are high-performance insu-

5.2 建筑中应用太阳能
5.2 Application of Solar Energy in Buildings

lation. This example is a thermocompression solar heater. At the bottom of the water tank there are pipes connecting with the room for supplying hot water and compensating cold water.

彩图 5-3 为安装在屋顶的大面积太阳能热水器。

图 5-4 是安装在各家阳台上的太阳能热水器。屋顶日照好，但高层建筑的下层住户管道长，投资大，且水温降低多。

太阳能热水器安装在阳台上好处多：投资少，管理方便，尤其当太阳能热水器与阳台栏板结合为双功能（防护与供热水）构件时则更先进。

Color picture 5-3 shows the solar heater on roof.

Fig. 5-4 shows solar heaters installed on each home's balcony. Sunshine is better on roof. But for the homes at lower parts of high rise buildings the pipes to the heaters are longer. Hence the cost is higher but the water temperature is reduced further.

To install solar heater on balcony can reduce investment and manage easily. It is especially advantageous when solar heater is integrated with balcony balustrade to become a double-duty element(protection & supplying hot water).

图 5-4　太阳能热水器安装在阳台上

Fig. 5-4　Solar heaters installed on balconies

(2) 玻璃真空管热水器

这种热水器目前国内外采用最广泛。图 5-5(a) 为其构造及支架图（作者研究：更先进的构造应该是真空管集热器与建筑构件，如屋面板或阳台栏板相结合的多功能构件）。

玻璃真空管集热器，由两层玻璃管组成，中间夹一层真空，消除对流失热。外层玻璃为高透明耐腐蚀、耐冰雹冲击的玻璃管。内层玻璃管外表有选择性镍黑涂层。该系统比金属管热水器优越。

(2) Glass vacuum solar heater

This type of solar heater is the most popular type used at home and abroad. Fig. 5-5(a) is its details of construction and the support frames. (According to the author's research, a more advanced construction is to combine the glass vacuum heater with roof panel or balcony balustrade to form a multi-function element).

A vacuum-glass tube heater is composed of double layers of glass tubes with a vacuum gap between them to avoid convective heat loss. The outer glass tube is high transparent, corrosion-proof and hail impact resistant. The inner glass tube's outer surface is painted by selective nickel black paint. This system is better than the metal tube heater.

5 太阳能与沼气在建筑中的应用
5 THE APPLICATION OF SOLAR ENERGY AND METHANE IN BUILDINGS

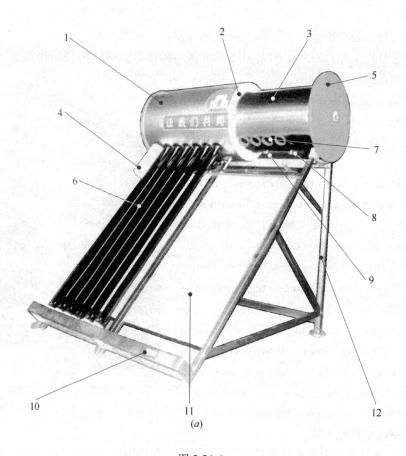

图 5-5(a)
Fig. 5-5(a)
(a)例：玻璃真空管太阳能集热器构造与支架
(a)An example: construction detail and support frame of a glass-vacuum-tube solar collector

1—水箱外壳：0.5mm 氧化铝板 Out shell: 0.5mm anodized aluminum plate
2—保温层：45mm 聚氨酯整体发泡 Insulation: 45mm polyurethane
3—水箱内胆：0.6mm 不锈钢板 Water tank: 0.6mm stainless steel plate
4—电加热器：保证阴雨天气仍能获得热水。Electric heater: hot water supply during non sunny days
5—水箱外端盖：工程塑料压注成型 Water tank end cover: plastic board
6—玻璃真空管集热管 Glass-vacuum collector
7—硅密封胶圈 Silica gel sealing washer
8—不锈钢管接头 Stainless connector
9—水温水位仪接口 The socket for water temperature & level meter
10—产品编号 Product number
11—反射板：0.5mm 防腐铝板 Reflective plate: 0.5mm anodized aluminum plate
12—支架 Support frames

5.2 建筑中应用太阳能
5.2 Application of Solar Energy in Buildings

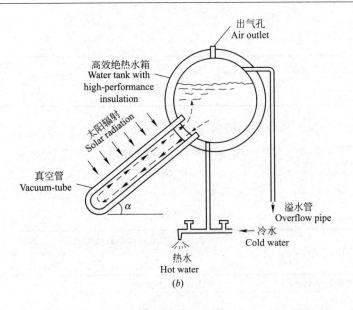

图 5-5(b)

Fig. 5-5(b)

(b)玻璃真空管热水器运行原理

(b) The operative principle of glass-vacuum tube solar collector

图 5-6　清华阳光集团售与拉萨一用户的
玻璃真空管热水器组

Fig. 5-6　A group of glass-vacuum-tube solar heaters
in Lhasa sold by Qinghua Sunlight Group

表 5-2(a)为该系统在西安(太阳能条件中下)平均日照下月均出水温度。

Table 5-2(a) shows the system's monthly mean temperatures of the output water under average sunshine condition in Xi'an (Solar energy condition is below average).

5 THE APPLICATION OF SOLAR ENERGY AND METHANE IN BUILDINGS

表 5-2(a)
Table 5-2(a)

月份 Months	1	2	3	4	5	6	7	8	9	10	11	12
水温(℃) Water Temp(℃)	33	33	36	42	49	55	55	55	40	35	32	30

表 5-2(b)为三种热水器比较根据清华阳光集团资料

Table 5-2(b) shows a comparison among three types' heaters (source: Qinghua Sunlight Group's data).

表 5-2(b)
Table 5-2(b)

类别项目 Types Items	太阳能热水器 RMB￥ Solar heater (100L, 1.5m² area)	液化煤气热水器 RMB￥ LCG heater (￥45.00/bottle)	电热水器 RMB￥ Electric heater (￥0.40/kWh)
投资 Investment	￥2400.00	700.00	950.00
使用寿命(年) life span(year)	15	6	6
每年使用天数 working days/year	300	300	300
每年动力费 Power cost/year	￥0	￥620	￥638
15年总用费 Total cost/15years	￥2400	￥11300	￥11920
安全事故 Safety Incidents	未发生 No	发生过 Occurred	发生过 Occurred
环境污染 Environmental pollution	无 No	有 Existed	有 Existed

图 5-7 为安装在坡顶上的大型玻璃真空管太阳能热水器系统（管外径110mm），新建或现有房屋平、坡顶均可安装。

Fig. 5-7 shows a large vacuum glass tube solar heater (outer diameter 110mm) system installed on a pitched roof. This system can be installed on flat or pitched (sloping) roof of new building or the existing.

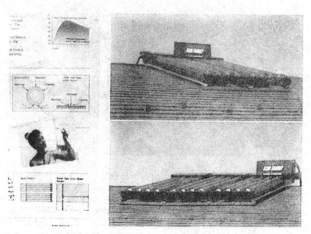

图 5-7 大型玻璃真空管太阳能热水器安装在坡顶上（日本）
Fig. 5-7 A large vacuum glass tube solar heater installed on the pitched roof (Japan)

(3) 热管式集热器

图 5-8 为一新型集热器——热管式集热器。图 5-9 为其运行原理图。

5.2 建筑中应用太阳能
5.2 Application of Solar Energy in Buildings

(3) Thermal-tube collector

Fig. 5-8 shows a new type of collector-thermal tube collector.
Fig. 5-9 shows its operational principle.

图 5-8 热管式集热器实验例

Fig. 5-8 A test example of a thermal-tube collector

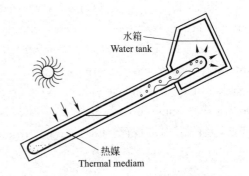

图 5-9 热管式集热器运行原理

Fig. 5-9 Working principle of thermal-tube collector

热管式集热器的工质是一种相变材料，密封于管内，管外表及底板外表均有选择性涂层。工质受到太阳辐射时，升温、蒸发（工质为液体），或升华（工质为固体），将热量带入水箱加热贮水，工质被冷却，恢复液态或固态，受重力作用返回低端，再受热再重复上述运行。

The working medium of thermal-tube collector is a phase-change material sealed in the tubes. The outer surfaces of the tubes and their underpan are painted by the selective paint. Under solar radiation, the working medium raises temperatures and evaporates (when working medium is a liquid) or sublimates (when working medium is a solid) to carry the heat into the water tank and conduct the heat to the water. When the medium is cooled and returned into liquid or solid form, it then returns to the low end by gravity and is heated again by the solar radiation and repeats the above process.

(4) 空心黑陶瓷板集热器

我国山东率先发展了多种类型空心黑陶瓷热水集热板，外形似多孔钢筋混凝土（RC）楼板，空心孔内冲水，向阳面覆透明盖板。实验、实例均证明效率高，取材易，造价低，耐久，很有发展潜力。

(4) Hollow black ceramic solar collector

Various forms of hollow black ceramic solar collectors have been firstly developed in Shandong Province, China. Their shapes are similar to reinforced concrete floor slabs with multiple-holes. Instead its holes are filled with water and the sunny surface installed with a transparent cover. Experiments and examples have proved that they are efficient, low cost, durable and using ordinary materials and have great potential.

(5) 太阳能灶

图 5-10 为日本展出的透过（或折射）聚焦太阳能灶，放在地面或多层建筑向阳阳台上均可用。

5 太阳能与沼气在建筑中的应用
5 THE APPLICATION OF SOLAR ENERGY AND METHANE IN BUILDINGS

图 5-11 为展示者用该太阳灶煎荷包蛋(日本神户世界太阳能大会上的演示)。图 5-12 为中国农村采用的反射聚焦太阳灶,用钢丝网水泥砂浆胚壳贴玻璃片制成,取材易、制作便、造价低,并有较好的效率(采光面积 2.5m² 焦斑区温度可达 680℃)。

(5) Solar cooker

Fig. 5-10 shows a transmissive(or refractive) focus solar cooker exhibited in Japan. It suits to be installed on both the ground and sunny facing balconies of multi-story buildings.

Fig. 5-11 the exhibitor is cooking fried eggs with the cooker(A product demonstration at the Solar World Congress, Kobe, Japan). Fig. 5-12 shows the reflective focus solar cookers used in Chinese villages. This type of cooker can be made of a wire reinforced concrete shell adhered with glass pieces. It is easy to use ordinary materials, easy to make and low cost and has better efficiency(with solar collective area of 2.5m², the temperature can reach 680℃ in focus zone).

图 5-10 透过(或折射)聚焦太阳能灶
Fig. 5-10 Transmissive(or refractive) focus solar cooker

图 5-11 透过聚焦太阳灶煎荷包蛋
Fig. 5-11 Transmissive focus solar cooker cooking eggs(fried eggs)

临西水泥太阳灶

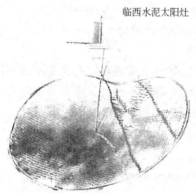

虞城水泥薄壳太阳灶

图 5-12 例:中国农村用反射聚焦太阳灶
Fig. 5-12 Examples: Reflective focus solar cookers used in Chinese villages

5.2.2 中国古代太阳能集热器

我国祖先早在2000多年前就曾发明削冰制成透过聚焦的透镜收集太阳能点火,古

5.2 建筑中应用太阳能
5.2 Application of Solar Energy in Buildings

书❶有载：西汉时古人"削冰令圆，举以向日，以艾承其影，则火生"（艾是一种易燃药用植物；"影"即焦点）。

图 5-13 为汉代与唐代的阳燧，即太阳能集热器。2000 多年前的西汉，我们祖先就削冰成透镜做成透过聚焦太阳能集热器；汉朝和 1000 多年前唐朝就能制成金铜合金反射聚焦凹镜太阳能集热器，确实反映了我们祖先的聪明才智。深深遗憾的是，由于历代王朝对科技的不重视，致使这些先进的科技成果惜遭夭折。

发展太阳能热水器与太阳能灶有利于环保，节能，节财。厨房与阳台连接朝向太阳对应用太阳能热水器与太阳能灶会创造更好的条件，并会大大改善厨房家务条件。

5.2.2 The ancient Chinese solar collectors

More than 2000 years ago, our ancestor already invented transmissive focusing lens of cut ice to collect solar energy for making a fire. An ancient book❶ has documented: In West Han Dynasty ancestors "cut ice to form a round lens, lift up to facing the sun, put a moxa in the shadow (the focus zone), the moxa is burning" (Moxa is a combustible medicine plant; shadow is the focus).

Fig. 5-13 shows a Yangxi, i. e. the solar collectors in Han Dynasty and Tang Dynasty. More than 2000 years ago in Western Han Dynasty, our ancestors had cut ice to form transmissive focusing lenses as solar collectors. There were reflective focusing concave lenses as solar collectors made of alloy of half gold and half copper in Han Dynasty and Tang Dynasty in more than 1000 years ago. These do reflect our ancestors' intelligence and wisdom. It's a deep pity that these advanced science and technology development were regrettably died young because of the negligence from the past governments.

图 5-13 中国古代反射聚集太阳能集热器例，左边的是汉朝的；右边的是唐朝的

Fig. 5-13 Example: Ancient Chinese reflective focusing solar collectors: left one is of Han Dynasty; right one Tang Dynasty

Developing solar heaters and solar cookers greatly benefits environmental protection and energy saving and money saving. Kitchen joining balcony facing the sun will be convenient to use solar hot water and solar cooker and improve the conditions of housework in the kitchen.

5.2.3 多功能构件

图 5-14 为屋面板与太阳能集热器相结合（正在安装中，日本）；图 5-15 为阳台栏板与集热器相结合（已有实验，中国）；图 5-16 为集热器与坡屋面相结合（英国）；图 5-17 为坡屋面结合太阳能集热器与太阳能电池板。

---

❶ 《问经堂丛书》(清·孙星衍·孙冯翼辑，十八种三十一卷)引，〈淮南 毕述〉载文. 西北大学历史系教授张永禄及科技史教授姚远提供史料.

❶ Source: Vol 31, Series 18 (Huainan, Bishu), Wenjing Tang Collections (Qing Dynasty, Editors: Sun Xingyan, Sun Fengyi), recommended by Professor in History From Northwest University: Zhang Yonglu and Professor in History of Science and Techology from Northwest University: Yao Yuan.

5 THE APPLICATION OF SOLAR ENERGY AND METHANE IN BUILDINGS

5.2.3 Multi-function members

Fig. 5-14 shows the combination of roof panel and solar collector (under construction in Japan). Fig. 5-15 shows a combination of balcony balustrade and solar collector (it has been experimented in China). Fig. 5-16 shows a combination of solar collectors and a pitched roof in England. Fig. 5-17 demonstrates a combination of sloping roof, solar collector and solar cell panel.

Fig. 5-14 Integrating roof panels with solar collectors (under construction in Japan)

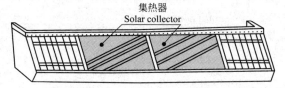

Fig. 5-15 Combining balcony railing panel with solar collector (experimented, China)

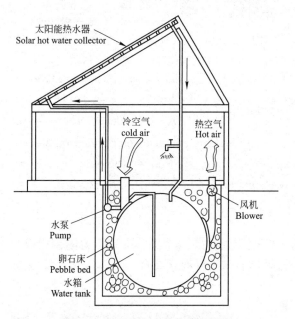

Fig. 5-16 The section of an experimental house of combining sloping roof with solar collector (UK)

Fig. 5-17 An example of a pitched roof integrated with solar heater and solar cell panel.

5.2 建筑中应用太阳能

5.2 Application of Solar Energy in Buildings

5.2.4 太阳能采光与日照

采光利用的是太阳有光感的波段380～730nm，日照波段更宽含红外线和紫外线。适当日照有消毒、杀菌、有利健康的功能。

(1) 透明屋顶大空间采光

彩图5-18为美国明尼苏达州首府明尼阿波利斯室内棒球运动场太阳能采光（自然采光）情况。天然照度系数达到8%，比我国一级采光天然照度系数高1%，是世界有名的大跨度（>100m）轻结构（双层透明薄膜充气屋顶）节能建筑。图5-19为其外观。

5.2.4 Day lighting and sunshine

Day lighting uses the visible waves between 380～730nm. Sunshine uses wider waves including infrared rays and ultraviolet rays. An appropriate exposure to the sun has benefits of disinfecting, killing germs to improve health.

(1) Large space lighting by transparent sun roof

Color picture 5-18 shows an indoor baseball field using natural day lighting in Minneapolis, MINNESOTA, USA. Its natural illumination factor is 8%, 1% higher than first grade of natural lighting in Chinese standard. It's a world renowned large span(>100m) energy saving building with light structure(double transparent plastic film inflated roof). Fig. 5-19 shows its external view.

图5-19 美国明尼苏达州，明尼阿波利斯棒球场双层透明塑料膜充气屋顶

Fig. 5-19 A baseball field with inflated roof of double transparent plastic film in Minneapolis, Minnesota, USA.

(2) 环形带窗大空间采光

(3) 地下空间太阳能采光

实例：

1) 下沉式庭院

图5-21为下沉式庭院自然采光，是联合国教科文组织设在巴黎总部的两层地下建筑群（共六座下沉式庭院）。中国下沉式窑洞采光也是如此。

2) 带形天窗

图5-20 为日本一环形带窗采光例。彩图5-24为巴黎地下超级市场所用带形天窗采光。

3) 星点顶光天窗

彩图5-25为法国HAGETMAU市政府（设在地下）会议室星点顶光天窗采光。

(2) Large space lighting with circular stripe windows

(3) Natural day lighting for underground space

For examples:

1) Sunken courtyard

5 THE APPLICATION OF SOLAR ENERGY AND METHANE IN BUILDINGS

Fig. 5-21 shows the application of day lighting in sunken courtyard of the UNESCO Headquarters, Paris. They are used for the 2-story underground office buildings in total six courtyards. Day lighting application in Chinese sunken caves' is the same.

2) Stripe skylights

Fig. 5-20 An example of circular stripe windows for day lighting, Japan. Color picture 5-24 shows the stripe skylight day lighting for the underground supermarket, Paris.

3) Nature star skylights

Color picture 5-25 shows day lighting with star skylights at the meeting room in HAGETMAU City Government building (built underground), France.

图 5-20 日本神户某运动场多层环形带窗采光，顶部天窗可采光同时通风

Fig. 5-20 Multi-tiered circular stripe windows for day lighting, the top skylight for lighting also act as ventilation in a sports ground, Kobe, Japan.

图 5-21 巴黎联合国教科文组织总部两层的下沉式办公楼群下沉庭院自然采光

Fig. 5-21 Day lighting in sunken courtyard for the 2-story sunken courtyard office buildings in UNESCO Headquarters, Paris.

（4）几何光学引光引景系统

图 5-26 为美国明尼苏达大学地下实验室中心采用的几何光学引光引景系统。该系统可将阳光与地面景引入地下房间。

图 5-27 为上述引光引景塔楼实景。

引光系统有双控装置：一是光敏镜头引导捕光镜直视太阳；二是电脑计算太阳轨迹，两者同时确认，捕光器才会运转，可避免如下误导：当太阳下有云层遮挡，而太阳对面又有云层反射光，光敏镜头误认该反射光为直射光欲使捕光器偏转，但电脑不确认，并制止误运转。

引景系统由反射镜与透镜组成，可将地面动、静景色引入地下房间，减轻或消除身居地下的忧闷感。

(4) Geometrical optical light and view-catcher system

Fig. 5-26 shows the geometrical optical light and view-catcher system of the underground laboratory centre in Minnesota University. This system can guide the sunlight and aboveground landscape into underground rooms.

Fig. 5-27 shows the real sight of above light & view-catcher tower.

The light guiding system has a double control. One is a light-sensitive lens leading the

5.2 Application of Solar Energy in Buildings

light catcher to look at the sun directly. The other is a computer to calculate the sun's track. Only when both control confirm the location at the same time, the catcher operates. This can avoid misleading the catcher under the following condition: when a cloud is under the sun to block the sunlight while a cloud opposite the sun reflects the sunlight, the lens might wrongly perceive the reflecting light as the sun's direct light and order the catcher to operate. But if the computer does not confirm, it can stop the wrong operation.

The view guiding system consists of reflective mirrors and lenses. It can guide the moving or still sights on the ground into underground rooms so as to reduce or eliminate the depressing feeling of living underground.

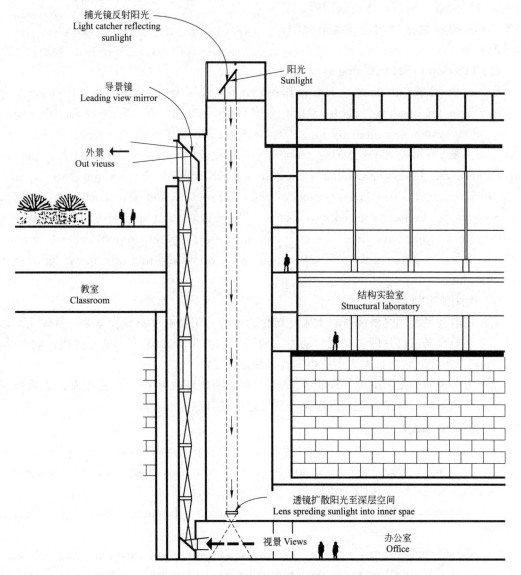

Fig. 5-26 The geometrical optical light & view-catcher system used in the underground laboratory centre Minnesota University. (Source: Dongnan Univ Master Wuhong)

5 THE APPLICATION OF SOLAR ENERGY AND METHANE IN BUILDINGS

(5) 光导纤维引光系统

光导纤维引光是靠光波在光导管内全反射，将光波从输入端传至输出端。图 5-28 为日本在神户世界太阳能大会展出的光导纤维引光系统。该系统由三部分组成：输入端为喇叭形捕光器，对准太阳，将阳光集束进入光导管（由光导纤维与护管组成）。该管可以弯曲穿引进入房间。每一光导纤维管终端有一小光点。这些小光点可编排成圆形或线形或平面形，外覆透明罩，于是我们即可获得圆形照明，或线形照明，或面形照明。

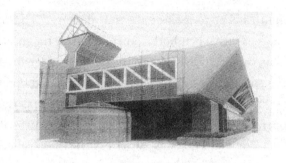

图 5-27 上述引光引景塔楼实景；右侧为塑料管盛水的水墙（太阳能供暖）

Fig. 5-27 A real sight of the above light & view-catcher tower; the right side showing a water wall made of plastic tubes filled with water(solar heating)

(5) Fiber optic light guiding system.

Fiber optic light guiding system is based on the light waves can travel from input of the fiber light guiding tube wholly reflectively to the output. Fig. 5-28 is the fiber optic light guiding system exhibited by Japan at the Solar World Congress, Kobe Japan. This system is composed of three parts: trumpet-shaped light catcher, guiding tubes and the lighting outlets. The light catcher aiming at the sun collects the sunlight into the fiber optic guiding tube, which is composed of the fiber optics, and a protective tube. The guiding tube can be bent and led into rooms. There is a small spot of lighting output at the end of each fiber optic line. We may lay out these spots into round, linear or a plane shape with a transparent cover. Then we can get a round lighting or a linear lighting or plane lighting.

(6) 太阳能电池

太阳能电池是利用光敏元件将太阳光转化为电能，用蓄电设备储存下来，从输出端用导线接至用电设备或灯具即可得到电动力或照明。彩图 5-22 德国太阳能电池板，约 $9m^2$，1kW 功率。彩图 5-23 为日本太阳能电池瓦每块功率 2W。

以上为非透明的太阳电池。现正在发展一种透明太阳电池板，可装在窗户上或屋顶上，尤其适宜用在可种植温室，既可采光又可得电和日照。

(6) Solar cells

Solar cell is a photo sensor to convert solar light into electricity stored in a storage battery. With conductor to joint the output and electrical appliances or lamps we can get power or lighting directly. Color picture 5-22 is a German solar cell panel about $9m^2$, 1kW. Color picture 5-23 shows a Japanese solar cell tile, each tile 2W.

The above are (opaque) non-transparent solar cells. Now, another solar cell panel, transparent solar cell panel, is being developed. It can be installed on window or roof, especially for plantable greenhouse. We can receive natural lighting, electricity and sunshine.

5.2 建筑中应用太阳能
5.2 Application of Solar Energy in Buildings

(7) 日照和太阳卡

现介绍一种运用太阳轨迹图(太阳卡)可便捷地得出:日出日落时间,昼夜时数,任一时刻太阳高度角,方位角,室内日照面积,室内最大可能日照时间与范围,建筑阴影等量值。

图 5-29(a)是以地理纬度北纬 36°作出的太阳卡,适用于地理纬度 34°~38°地区,如西安、兰州等。

图 5-29(b)为不同纬度的三幅太阳卡供选用。南半球相当地区可按对称原理使用。先了解太阳卡:太阳卡图中,9 个同心圆为太阳高度角环(太阳高度角:以太阳作为一个点与地面观察点之间的连线与过该观察点的水平面之间的夹角,参见图 5-30)。每两相邻同心环之间的高度角之差为 10°。当太阳正处于观察点正上方(天顶)时,该高度角(即圆心所示高度角值)为 90°;过圆心的 18 条放射线为方位角线。(方位角:参见图 5-30,以该正南线为始线,当地太阳时正午时刻,方位角为"零"),上午方位角为逆时针向为"负";下午为顺时针向,方位角为"正"。图中标有 4、5、6、7、8、9、11、13、15、16、17、18、19 数字的曲线(太阳时基准线)及其辐射线组为太阳时线系。7 条粗黑线为所标节气日太阳空间轨迹在水平面的投影。

(7) Sunshine and solar chart

Now we'll introduce a solar locus picture(solar chart) to you. Using it we can easy calculate sunrise or sunset time(hour), daytime/night time hours, solar altitude angle and azimuth at anytime, indoor sunshine area, indoor possible maximum sunshine hours and area, building shadow area, and etc.

Fig. 5-29(a) is a solar chart made for geographic location at latitude $\varphi=36°N$. It's suitable to areas between $\varphi=34°$ and $38°N$, such as Xi'an and Lanzhou, etc.

Fig. 5-29(b) shows three available solar charts for different latitudes. For the corresponding places in the Southern Hemisphere these charts can be used with symmetrical principle. First let's see how to understand solar chart. In a solar chart there are 9 concentric circles which are the circles of solar altitude angles(Solar altitude angle: taking the san as a point, the solar altitade angle is the angle betuieen the line of the sun to the viewpoint and the plane through the viewpoint, see Fig. 5-30). Between any two adjacent concentric circles the altitude angle difference is 10°. When the sun is just up the viewpoint(Zenith) the altitude angle(i. e. the altitude angle of centre) is 90°. Through the centre the 18 radiate lines are azimuth lines(Azimuth, see Fig. 5-30, is taking due south line as the starting line, at the local noon of solar hour the azimuth is "zero"). Before noon azimuths are "negative" anti-clockwise. After noon azimuths are "positive" clockwise. In the solar chart the curve marked 4、5、6、7、8、9、11、13、15、16、17、18、19(solar hour fiducial line), with its radiate lines is the solar hour line system. The 7 thick-black lines are the projections on the horizontal plane of the marked Solar Terms' solar locus in the sky.

现以西安为例说明太阳卡的应用:

1) 求西安地区夏至、春分、秋分、冬至日出、日落时间及昼夜时间。

[解]:西安地理纬度 $\varphi=34°16'N$,适用太阳卡 $\varphi==36°N$(图 5-29(a)):

5 THE APPLICATION OF SOLAR ENERGY AND METHANE IN BUILDINGS

图 5-28 日本神户世界太阳能大会，
日本展出的光导纤维引光系统

Fig. 5-28 The fiber optic light guiding system exhibited by Japan at the Solar World Congress, Kobe, Japan

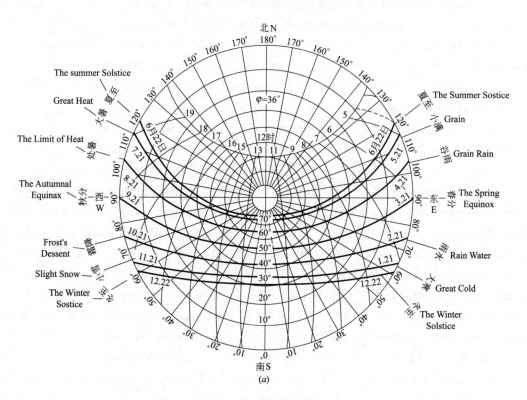

图 5-29(a)

(a)北纬 $\varphi=36°$ 太阳卡(适用于 $\varphi=34°$ 至 $38°$ 地区)

Fig. 5-29(a)

(a) Solar chart $\varphi=36°N$ (suitable to areas between $\varphi=34°$ and $38°$ latitude)

5.2 建筑中应用太阳能
5.2 Application of Solar Energy in Buildings

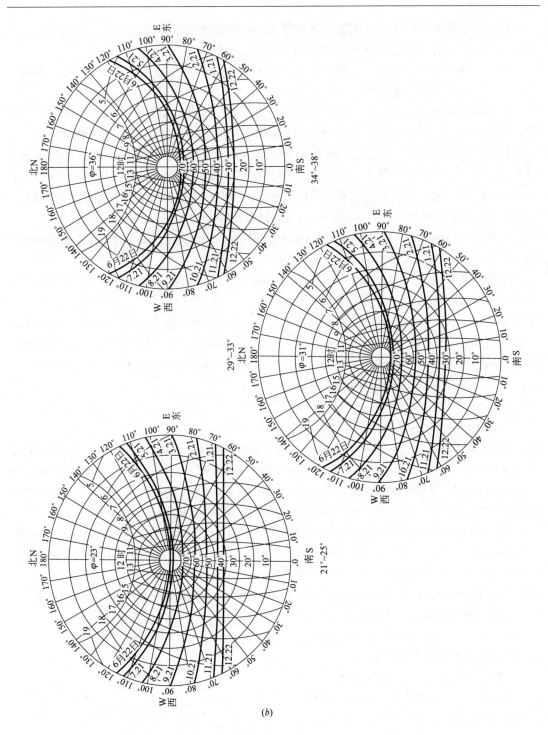

(b)

图 5-29(b)

(b)三种不同高度角太阳卡：$\varphi=23°\text{N}$(适用于 $\varphi=21°\sim25°\text{N}$ 区)，
$\varphi=31°\text{N}$(适用于 $\varphi=29°\sim33°\text{N}$ 区)，$\varphi=36°\text{N}$(适用于 $\varphi=34°$ 至 $38°\text{N}$ 区)

Fig. 5-29(b)

(b)Three different latitudes' Solar Charts: $\varphi=23°\text{N}$ (suitable to $\varphi=21°$ to $25°\text{N}$ latitudes);
$\varphi=31°\text{N}$ (suitable to $\varphi=29°$ to $33°\text{N}$ latitudes); $\varphi=36°\text{N}$ (suitable to $\varphi=34°$ to $38°\text{N}$)

5 太阳能与沼气在建筑中的应用
5 THE APPLICATION OF SOLAR ENERGY AND METHANE IN BUILDINGS

从太阳卡夏至太阳轨迹线起始端按相邻太阳时辐射曲线作相似虚曲线，交太阳时基线于4～5之间一点，可估出4：40，即该日出时间。

24－4：40＝19：20 即该日落时刻。白天＝(19：20－4：40)＝14 小时 40 分，夜晚＝24：00－14：40＝9 小时 20 分。

从太阳卡春分、秋分线始、末端看出，它们分别交于太阳时线的辐射线 6：00 与 18：00 处，表明春分、秋分日出时间为 6：00，日落时间为 18：00，昼夜均为 12 小时。

从太阳卡冬至线起始端交太阳时线的辐射线"判断"日出时间为 7：20，日落时间为 24：00－7：20＝16：40，白天＝(16：40－7：20)＝9 小时 20 分，夜晚＝24：00－9：20＝14 小时 40 分。冬至夜晚的时间正是夏至的白天时间；冬至白天的时间正是夏至晚上的时间。

Now, let's take Xi'an as an example to explain how to use the solar chart.

1) To find the hours of the sunrise, sunset, daytime, night time, Summer Solstice, Spring Equinox, Autumn Equinox and Winter Solstice.

[Solution]: Xi'an geography latitude $\varphi=34°16'N$, the solar card of $=36°N$(Fig. 5-29 (a)) is suitable:

According to the adjacent solar hour radiate line, from the beginning of the Summer Solstice solar locus line on the solar chart to draw a similar dotted curve crosses the solar hour fiducial line at the point between 4-5 we may estimate 4：40 as the sunrise hour.

24-4：40＝19：20 as the sunset hour

Daytime＝(19：20-4：40)＝14h 40m

Nighttime＝24：00-14：40＝9h 20m.

From the solar chart's Spring Equinox and Autumn Equinox line we can see the line's beginning and end crossing the solar hour radiate curves at 6：00 and 18：00 points respectively. It shows that on the Spring Equinox and Autumnal Equinox sunrise time is at 6：00 and sunset time is at 18：00. The length of daytime and night time are both 12 hours.

On the solar chart, from the radiate curve of the Winter Solstice line, both the start and the end, crossing the solar hour line, we may estimate the sunrise is at 7：20am and sunset at 24：00－7：20＝16：40, daytime＝16：40-7：20＝9h 20m, night time＝24：00－9：20＝14h 40m. Winter night time hours＝summer daytime hours and winter daytime hours＝summer night time hours.

2) 求西安地区夏至、冬至 7：00、10：00、12：00、14：00、17：00 各时刻当地太阳时的太阳高度角 H 与方位角 A。

[解]：图 5-29(a)太阳卡仍适用。

找到 7：00 太阳时线辐射线与夏至太阳轨迹线交点作太阳卡圆心与该交点连线沿长交最外环一点，可判断出该方位角为 $A=-103°$；同时可看出 7：00 太阳时线辐射线与上述连线的交点正处于太阳高度角环 20°～30°之间，可判断出 $H=25°$。由对称原理可知：17：00 $A=$

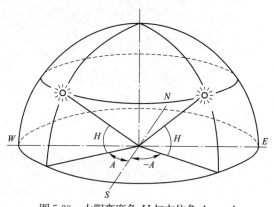

图 5-30 太阳高度角 H 与方位角 A；$-A$

Fig. 5-30 Solar altitude angles H and azimuths A；$-A$.

5.2 建筑中应用太阳能
5.2 Application of Solar Energy in Buildings

$103°$ $H=25°$,同样:$10:00$ $A=-75°$,$H=60°$,$12:00$ $A=0°$,$H=77°$,$14:00$ $A=75°$,$H=60°$。

冬至太阳时线 $7:00$ 的辐射线与太阳轨迹线无交点表明太阳还未出现;同理表明冬至 $17:00$ 太阳此前早已落山。同法得出冬至:

$10:00$ $A=-31°$,$H=24°$,$14:00$ $A=31°$,$H=24°$,$12:00$ $A=0°$,$H=31°$。

2) To calculate the solar altitude angles H and the azimuths A at the local solar hours: $7:00$, $10:00$, $12:00$, $14:00$, $17:00$ of Xi'an Summer Solstice and Winter Solstice.

[Solution]: Fig. 5-29(a) solar chart is suitable here too. Find the intersection point of $7:00$ solar hour radiate line and the Summer Solstice locus line, draw a joint line through the chart centre and the intersection point, prolong the line to cross the most outside circle at a point, we may read the azimuth $A=-103°$ and we can see the intersection point of the $7:00$ solar hour radiate line and the above joint line is just between the solar altitude angles' circles $20°$ to $30°$. We may read the altitude angle $H=25°$ according to the symmetric principle and read $17:00$ $A=103°$, $H=25°$, and $10:00$ $A=-75°$, $H=60°$, $12:00$ $A=0°$, $H=77°$, $14:00$ $A=75°$, $H=60°$.

We may notice that there is not any intersection point of the Winter Solstice solar hour $7:00$ radiate line and the solar locus line. It means that the sun has not appeared at this hour. Based on the same principle, we know that the sun has been already down before $17:00$.

With the same method we can get:
Winter Solstice:
$10:00$ $A=-31°$, $H=24°$, $14:00$ $A=31°$, $H=24°$, $12:00$ $A=0°$, $H=31°$

3) 求西安冬至 $12:00$(太阳时)室内日照面积。

[解]:按一定比例如 $1:100$ 绘出平剖面,如图 5-31。在平面图上靠窗两侧各作一条方位角 $A=0$ 的直线。在剖面图上按西安冬至正午太阳高度角 $H=31°$ 通过该窗上口外沿"e"点和窗下口内沿"f"点作两条阳光入射线,交地面于"g"、"h"点。由"g"、"h"分别引垂线交平面图 $A=0$ 的两直线于"a"、"b"、"c"、"d"4点,所围面积 abcd 即所求室内日照面积。

3) To find the indoor sunshine area at $12:00$(solar hour) on the Winter Solstice, Xi'an

[Solution]: Take a scale such as $1:100$ to draw a plan and a section as Fig. 5-31. On the plan against both sides of the window, make two straight lines of azimuth $A=0$. Draw two incoming sunlight lines separately through the window top outside point "e" and the window bottom inside point "f", according to the solar altitude angle $H=31°$($12:00$ solar hour, Xi'an). The two lines reached the ground at points "g" and "h". From "g" and "h" draw two vertical lines separately to cross the two straight lines of $A=0$ at points "a" "b" "c" "d". Then the enclosed area abcd is the indoor sunshine area.

4) 求西安冬至室内最大可能累积日照时间和累积日照面积。

[解]:设如图 5-32 窗户已定,仍用 $\varphi=36°$ 太阳卡。将卡中心与窗口"A"点重合。方向,上为北,连 AC、并延长交太阳卡最外环于"E"。该点位置在太阳时线 6:40,太阳还未出现。7:20 日出,室内开始有日照,1号和2号线才是真的起始入

5 THE APPLICATION OF SOLAR ENERGY AND METHANE IN BUILDINGS

室阳光束。由对称性可知，直到 16：40 太阳落山时，3 号和 4 号线为最终入室阳光束。

我们可看出最大可能累积日照时间为 7：20～16：40，共 9 小时 20 分；最大可能日照累积面积范围为 1→4 线所包面积 J。

4) To find the indoor possible accumulative maximum sunshine hours and area in the Winter Solstice in Xi'an.

[Solution]: If the window has been designed as Fig. 5-32, $\varphi=36°$ solar chart is used too. Allow the chart centre overlap the point "A" at the window, up direction is north, join and prolong AC to cross the most outside circle at point "E". "E" is located at 6：40 of the solar hour line and indicates the sun has not appeared. At 7：20 sunrise, the room begins to receive sunshine. Line 1 and Line 2 are true sunshine rays into the room. From symmetric principle we know that till 16：40 sunset hour Line 3 and Line 4 are the final sunshine rays into room.

We can see that the possible accumulative maximum sunshine hours are altogether 9 hours and 20 minutes between 7：20 and 16：40. The possible accumulative sunshine area is the enclosed area by Line 1→4.

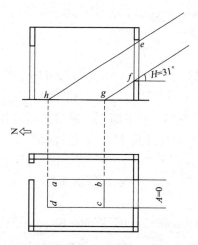

图 5-31 西安冬至 12：00（太阳时）室内日照面积

Fig. 5-31 Xi'an indoor sun shine area at the Winter Solstice 12：00(solar hour)

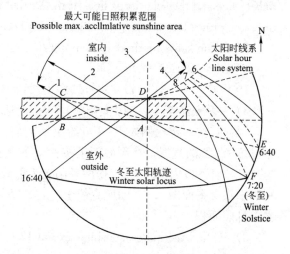

图 5-32 室内最大可能累积日照时间与面积

Fig. 5-32 Indoor possible maximum accumulative sunshine hours and area

5) 求夏至 10：00（太阳时）西安建筑阴影（图 5-33）。

[解]：按恰当比例（如 1：100）绘出该屋顶平面，标出 ABCDEFGH 诸点，将 DE 边与 $\varphi=36°$ 的太阳卡 NS（北南）轴重合，E 点与卡中心重合。由 10 点太阳时线的辐射线与夏至曲线的交点得太阳高度角 $H=60°$。自卡心与上述交点连线，延长交方位角环 $A=-75°$ 处。设房高 20m，无挑檐，自 "E" 点作垂线与方位角 $A=-75°$ 线正交。在此垂线上按同样比例取 $EK=20m$。自 "K" 点作 30°斜边，交方位角线于 E'，EE' 即 E 点外墙阳角的阴影，用轴测投影法可方便地作出该建筑的阴影，如图 5-33。

5.2 建筑中应用太阳能
5.2 Application of Solar Energy in Buildings

5) To find out building's shadow line at the Summer Solstice 10∶00(solar hour) in Xi'an. (Fig. 5-33)

[Solution]: Take a suitable scale(such as 1∶100) to draw the roof plan. Mark the points *ABCDEFGH* and let *DE* line overlap the NS axis in $\varphi=36°$ solar chart and overlap point "*E*" with the chart's centre. From the intersection point of the 10 o'clock solar hour radiate curve and the Summer Solstice curve we get the solar altitude angle $H=60°$. Join the chart's centre with the above intersection point and prolong it to cross the azimuth circle at $A=-75°$. If assume the building height$=20$m, without overhang, from point "*E*" draw a vertical line with a right angle to cross the line of azimuth $A=-75°$. On the vertical line with the same scale assuming $EK=20$m, from point "*K*" draw a hypotenuse with 30° angle to cross the azimuth line at point *E'*. *EE'* is indeed the shadow of the point E's external corner. Then with the axonometric projection method we may easily get the building's shadow as Fig. 5-33.

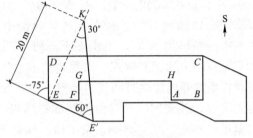

图 5-33 求建筑阴影
Fig. 5-33 To find out building's shadow line

5.2.5 太阳能建筑

所有建筑无不受到太阳能的恩惠。在此我们定义：经过良好设计，达到优化利用太阳能的建筑称太阳能建筑，以太阳能供暖分类，太阳能建筑分主动系统和被动系统两类：

(1) 主动系统

图 5-34 为主动系统的组成：集热器收集太阳热能，由工质(载热体：水或空气)将热量通过管道输入贮热体(工质为水时用水箱；工质为空气时用固体贮热物质如卵石)。贮热体经管道将载热体送入各散热器，以辐射与对流方式将热传给采暖房间。载热体散热后回到贮热体，再回到集热器被加热，水泵与风机分别作为水或空气运行的动力。

用本系统单纯供暖已越来越少(第一次投资高，又不全年使用)。

5.2.5 Solar buildings

No building has not benefit from solar energy. Here we define solar building as: buildings well designed to optimize the utilization of solar energy. There are two types of solar buildings with solar heating i.e. active system and passive system:

(1) Active system

Fig. 5-34 shows composition of an active system: collector collects solar heat; working fluid (heat carrier: water or air) transfers the heat into heat storage(water tank when water is the working fluid; or solid thermal mass such as pebbles when air is the working fluid). Heat storage medium transports the heat carrier by pipes into every radiator then transmits heat into rooms by radiation and convection. After release the heat into rooms the heat carrier returns back to the heat

5 THE APPLICATION OF SOLAR ENERGY AND METHANE IN BUILDINGS

storage then to the collector to receive heating again. Pump or blower is used respectively as the power to move the water or air.

Using this system for heating only is less and less (first investment is expensive. It is not used through whole year).

用本系统供热水及冬季兼供暖则日益受到重视。使用水为工质时,应注意水温与供热水量的优化关系。有人实验:用15℃的水加温到50℃,每平方米集热器出水60L;加热到60℃,出水30L。加热到70℃则只能出约15L热水。水温越高出水量越低。原因是水温高与外界温差大,失热多,需要在集热器内加热时间长。当然,所需水温多少还要满足生产或生活的要求。例如需要水蒸气,就采用回转弯曲长距离集热器,在输出端得水蒸气;还可在适当区段取得所需温度的热水。当需要热水量多时,则需大面积集热器。图5-35(a)为昆明某太阳能室内游泳池屋面大面积安装集热器。图5-35(b)为室内部分状况。该游泳池利用太阳能加热水是成功的,水温可保持所需24℃左右。但屋顶、外墙热阻太小,造成大量的内表面凝结水。

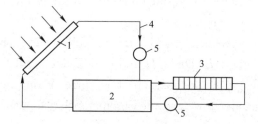

图 5-34 主动系统供暖
Fig. 5-34 Active system for heating
1—集热器 Collector; 2—储热箱 Water tank; 3—散热器 Radiator;
4—管道 Pipes; 5—水泵或风机 Pump or blower

彩图5-36为美国某公园露天太阳能游泳池。该地区尘沙少,设计者巧妙地将太阳能热水器组装成露天音乐场背景。此地已成为人们特别是年轻人喜欢的活动中心。

彩图5-37为法国HAGETMAU市地下热水露天游泳池。该市将深藏地下700m的高于60℃的热水用深水泵取出供全市生产、生活补充用水。本游泳池用水,则是将地热水多余热量取用后送入游泳池的水,水温23℃。

此外,如干燥室、预制构件养护室、多种类型工业、农业、民用温室等均可采用太阳能。太阳房、建筑绿化更是应用太阳能的广阔领域。

Using this system for supplying hot water and heating in winter has been increasingly popular. Particular attention should be paid to the optimal relation between the water temperature and the amount of hot water supply when using water as working fluid. A test has indicated that when water is heated from 15℃ to 50℃, hot water supply is 60L/m² collector; when heated to 60℃, the supply is 30L/m² collector; when heated to 70℃, only 15L/m² collector can be supplied. The higher water temperature required, the less hot water supplied. Because the higher water temperature demands more temperature difference between the water and the outside temperature. When heat loss is more, the water in collector should be heated for longer time. Of course the water temperature should satisfy the needs of production and living. For examples: when we need water vapor, we may use a long winding collector to get water vapor from the end. We can also collect hot water of certain temperature from a suitable position of the long collector. If large quantity of hot water is needed, we have to install large area of collectors. Fig. 5-35(a) shows many collectors installed on a large roof for an indoor swimming pool in Kunming. Fig. 5-35(b) shows a part of the inside. This swimming pool is successful in using solar ener-

5.2 建筑中应用太阳能
5.2 Application of Solar Energy in Buildings

gy to heat water and keep it at the required temperature at about 24℃. But the thermal resistance of the roof and exterior wall is very little and caused heavy condensation at the internal surfaces.

Color picture 5-36 is an outdoor solar swimming pool in a park, USA. There is little dust in the area. The designer cleverly installed the solar heaters as the background of an open music square. It has become a popular activity centre for people especially for youngsters.

Color picture 5-37 shows a swimming pool of geothermal water in HAGETMAU City, France. Geothermal water at 700m below ground with temperature higher than 60℃ has been drawn up by deep water pumps to supply the hot water for production and living. The left over geothermal water after the heat extraction is supplied to this swimming pool and remains at 23℃.

Furthermore, drying room, curing room for precast elements, various greenhouses for industry, agriculture and civil uses can all use solar energy. Solar buildings, landscaping of buildings are even wider fields for the application of solar energy.

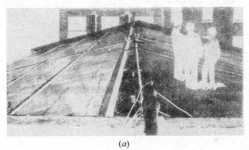
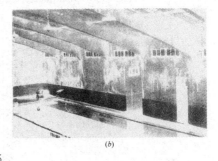

(a) (b)

图 5-35

(a)昆明某太阳能室内游泳池大面积屋面集热部分视景；(b)上述游泳池由于绝热差产生严重内表面凝结

Fig. 5-35

(a) A part view of the large area collectors installed on the roof of a solar indoor swimming pool, Kunming; (b) Poor insulation caused serious surface condensation in the above swimming pool

(2) 被动系统

被动式太阳房的特点是：将建筑物的全部或一部分既作为集热器，又作为贮热器和散热器；既不要连接管道；也不要水泵或风机。被动系统又分间接得热和直接得热。

1) 间接得热系统有特朗伯墙（Trombe wall）；水墙，载水墙（充水墙），和毗连日光间，毗连温室等。

(2) Passive system

The characters of passive solar buildings are the whole or part of a building acting as a collector as well as a heat store and radiator. No pipe, no pump or blower is needed. Passive system can be divided into indirect gain and direct gain systems.

1) Indirect gain systems have Trombe wall, water wall, water loaded wall, attached sunspace and attached greenhouse, and etc.

① Trombe wall 太阳房

图 5-38～图 5-42，为该太阳房冬日、夜和夏日、夜运行方式。

① Trombe wall solar house

Fig. 5-38, 39, 40, 41 and 42 shows the operating modes in winter day, night and summer day and night respectively.

5 THE APPLICATION OF SOLAR ENERGY AND METHANE IN BUILDINGS

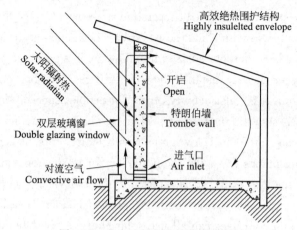

图 5-38 冬白天特朗伯墙太阳房供暖
Fig. 5-38 Winter daytime heating by Trombe wall solar house.

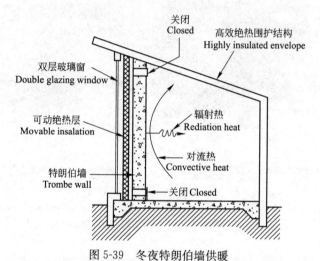

图 5-39 冬夜特朗伯墙供暖
Fig. 5-39 Winter night heating by Trombe wall solar house

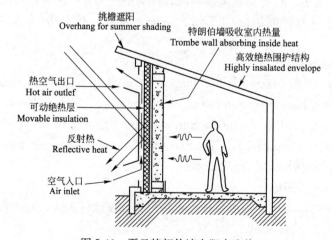

图 5-40 夏日特朗伯墙太阳房防热
Fig. 5-40 Summer daytime insulation by Trombe wall solar house

5.2 建筑中应用太阳能
5.2 Application of Solar Energy in Buildings

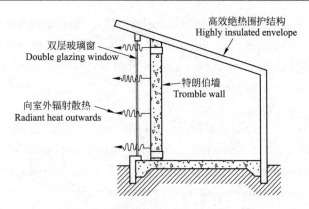

图 5-41　夏夜特朗伯墙太阳房防热
Fig. 5-41　Summer night insulation by Trombe wall solar house

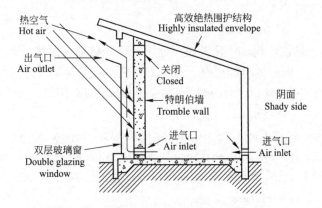

图 5-42　特朗伯墙太阳房夏季另一防热法
Fig. 5-42　Another summer insulation in Trombe wall solar house

冬日运行：

Trombe wall 太阳房主要创始人是法国教授 Felix Trombe，故称该集热墙为 Trombe Wall。其任务就是冬季白天吸热、贮热，晚上放热。Prof Trombe 最初实验时用 60cm 厚混凝土，但太厚，贮热容量太大，对室内达不到优化供热，后改为 40cm 厚就较满意了。

Trombe 墙上下有通风孔(图 5-38)。其外 8cm 空气层外侧装双玻窗，上下也有通风孔(图 5-40)。

冬季白天，靠空气层加热的空气对流加热房间。空气层的热能来自其温室效应。

冬夜运行：

图 5-39，冬夜由 Trombe 墙白天的贮热对室内以辐射与对流方式供热。为减少失热，除屋顶和外墙高绝热外，空气间层内设可动绝热层，冬季，白开夜闭，同时所有通风孔均关闭。

Winter daytime operation

The main inventor of Trombe wall solar building is a French Professor Felix Trombe. So the thermal mass wall is named as Trombe wall. Its duty in winter daytime is to absorb and store heat and release heat at night. Prof Trombe carried out their initial experiment with 60cm thick concrete wall. But the wall was too thick to optimally release heat into room because of too much heat storage capacity. Later, when wall is changed to 40cm thick the result is satisfactory.

Trombe wall has vents at the top and bottom(Fig. 5-38). A double-glazing window is installed outside its 8cm external air space. The air space also has top and bottom vents. (Fig. 5-40)

In winter daytime the heated air in the air space heats the room by convection. The heat energy in the air space is generated from greenhouse effect.

Winter night time operation

As shown in Fig. 5-39, during winter night, the stored heat in the Trombe wall heats the room by radiation and convection. Besides high-insulation of roof and exterior wall, adjustable insulation should be installed in the air space to reduce heat loss. In winter daytime it should be open and have all vents closed at night.

夏日运行：

图 5-40，为阻止进热：a)可动绝热层及 Trombe 墙上下通风孔关闭，阻止进热；b)玻璃上下通风孔开启散热。

夏夜运行：

图 5-41，可动绝热层收起，以便散热，通风孔可开启加强散热。有的地区如重庆，深夜仍有岩石辐射热，开窗室内可能更热。改善方法之一就是环境及建筑绿化。

夏季另一运行方式就是建筑物背阴面外墙底部开通风孔(图 5-42)。该处气温较低时可用。

Summer daytime operation

As shown in Fig. 5-40, there are two ways to stop heat gain. a) The adjustable insulation and the vents at the top and bottom of the Trombe wall should be closed. b) The vents at the top and bottom of the glass should be open to release heat by ventilation.

Summer night time operation

As shown in Fig. 5-41, the adjustable insulation should be open to release heat. Vents can be open to maximize the releasing of heat. In place such as Chongqing, late at night there still has radiant heat from rocks. Rooms might be even hotter if windows are open. One of the methods of improvement is to landscape the environment and buildings.

In summer another operation method is to install vents at the bottom of the wall on shady side(Fig. 5-42). The vents can be used when air temperature there is lower.

② 水墙

明尼苏达大学地下实验中心的地面部分(图 5-27)采用了 30cm 直径的塑料管盛水的水墙。图 5-43 为国外一水箱存水的水墙太阳房。

水吸热能力是混凝土、砖、土等材料的 4～5 倍，贮同量的热，水的质量只是上述材料的 1/4 或 1/5，这是水墙的一大优点。

③ 充水墙(载水墙)

水墙的缺点是时间延迟少，外侧受热后易通过对流传热到内侧，可能过早向室内散热，使室内白天产生过热。固体材料时间延迟较长，是因为内部无对流传热。充水墙就是用钢筋混凝土做外壳，里面充水的墙。这种墙有相当的热容量和时间延迟。

图 5-44 为国外实验的一种充水墙的温度曲线。该墙是一种 RC 空心墙板，壁厚 5cm，外形为 1.2m×2.4m×25cm，嵌入塑料管再充水，向阳面涂选择性黑色涂层。水墙及充水墙还有两个优点，一是对防火较有利；二是热季可取用孔内储了热的温水，以冷水充换，

5.2 建筑中应用太阳能
5.2 Application of Solar Energy in Buildings

同时使室温降低。

2007 年,美国一研究单位用循环水建了一幢水墙太阳房展出,并计划批量生产。

② Water wall

The aboveground part of the underground laboratory centre in Minnesota University, has used a water wall made of 30cm diameter plastic tubes containing water(see Fig. 5-27). Fig. 5-43 shows an example of water wall solar house with water tank containing water in a foreign country

The water's ability of absorbing heat is 4 to 5 times of concrete, brick and soil and etc. To store equal heat, the mass of water is only 1/4 or 1/5 mass of the above material. This is a great advantage of water wall.

③ Water loaded wall

The shortcoming of water wall is its short time delay, easy convection of outside heat to inside and its potential of overheating the room too early. Solid material has longer time delay because of no convection inside the mass. Water loaded wall is the wall with a reinforced concrete shell filled with water. It has a considerable heat capacity and time delay.

Fig. 5-44 shows the temperature curves from an experiment of water-loaded wall in a foreign country. This wall is a hollow reinforced concrete panel. The shell of the panel is 5cm thick and its external size is 1.2m×2.4m×25cm. Plastic tubes are set in the cavities and filled with water. The surface facing the sun is painted with black selective painting. Water wall and water loaded wall have two advantages: one is good fireproofing; the other is that in summer we can use the heated water stored in the cavity. It can reduce the room temperatures when warm water is replaced with fresh cool water.

In 2007 in US, a research institute has built and exhibited a water wall solar building with recycle water and planned to produce more in batches.

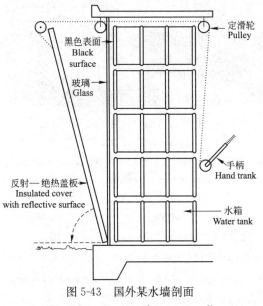

图 5-43 国外某水墙剖面

Fig. 5-43 Section of a water wall in a foreign country

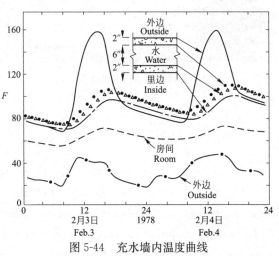

图 5-44 充水墙内温度曲线

Fig. 5-44 The temperature curves in a water loaded wall

5 太阳能与沼气在建筑中的应用
5 THE APPLICATION OF SOLAR ENERGY AND METHANE IN BUILDINGS

④ 毗连日光间和种植温室

图 5-45 为法国鲁伯教授家毗连日光间太阳房。玻璃内侧有半透明可动绝热层,用作减少夏进热,冬失热。日光间顶部有可控通风窗。

图 5-46 为甘肃自然能源研究所太阳能研究基地食堂毗连日光间。作者曾测到:室外 -9℃,毗连日光间内 12℃(有阳光),效果甚佳。

我们的毗连日光间节能实验:

图 5-47 为 2007 年 2 月 16~22 日(多云/阴转晴)我们的实验,测气温 40 次,阳台阳光间 13~28℃,平均为 23℃;主室 21~25℃(供暖),平均为 23.4℃;室外 7~22℃,平均为 15.4℃。若没有阳台阳光间,主室将按:

$\Delta t = 23.4 - 15.4 = 8K$ 失热。有阳台阳光间,则按:

$\Delta t = 23.4 - 23 = 0.4K$ 失热。证明,阳台阳光间有很高的节能效益。40 次测温中有 18 次阳台气温比主室气温高。其时,主室非但没有失热,反而会从阳台得热!有种植的温室效益更好,在光合作用下吸 CO_2 吐 O_2,更有利于节能减排,消除气候过暖。

④ Attached sunspace and planted greenhouse

Fig. 5-45 is an attached sunroom in the solar house of Prof LOUBE's home in France. A layer of semi-transparent controllable insulation is inside of the glazing. It can reduce summer heat gain and winter heat loss. Controllable vents have been installed on top of the sunroom.

Fig. 5-46 shows an attached sunroom of the dining room at Solar Study Base of Gansu Institute of Natural Energy. The author has measured: when outside was -9℃, inside the sunspace was 12℃. The effect was very good.

图 5-45 法国 LOUBE 教授家毗连
日光间太阳房

Fig. 5-45 The attached sunroom in solar house of Prof LOUBE's home, France

图 5-46 甘肃自然能源研究所太阳能研究
基地餐厅毗连日光间

Fig. 5-46 The attached sunspace of the dining room in Solar Study Base of Gansu institute of Natural Energy

We have carried out tests on the energy saving of attached sunspace:

Fig. 5-47 shows one test between Feb 16 and 22 2007. Among 40 measured air temperatures, the temperatures in balcony sunspace is between 13~28℃, mean of 23℃, main room 21~25℃ (heating), mean of 23.4℃, outdoor air 7~22℃, mean of 15.4℃. If there were no balcony sunspace, the main room would have lost heat according to the equation: $\Delta t = 23.4 - 15.4 = 8K$; With the balcony sunspace, main room will lose heat according to. $\Delta t = 23.4 - 23 = 0.4K$.

5.2 建筑中应用太阳能
5.2 Application of Solar Energy in Buildings

It proves that balcony sunspace has high efficiency of saving energy. In the 40 measurements, 18 measurements from the balcony were higher than those in the main room. The main room not only had no heat loss but indeed had heat gain from the balcony! Planted greenhouse has even better effect. Under photosynthesis, the plants absorb CO_2 and release O_2. It's more advantageous for energy saving, reducing CO_2 emission and slowing down climate over warming.

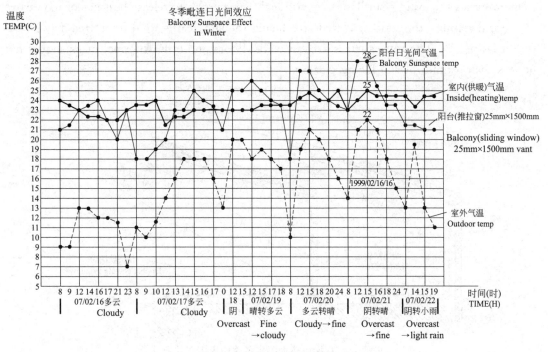

图 5-47 毗连日光间节能实验

Fig. 5-47 The energy saving test of an attached sunspace

2）直接得热系统

直接得热太阳房是通过透明构件（例如窗户）直接进入太阳辐射，由内部墙、楼板或底层地面、家具设备等作为贮热物质，白天吸热储热，夜间供暖。为减少失热，夜间应盖绝热窗帘，盖在冷侧，以避免或大大减轻玻璃内侧面的凝结水。绝热窗帘与窗户周边接缝气密性好，也可放在内侧，管理较方便。图 5-48、图 5-49 示出了直接得热太阳房冬季白天、夜间的工作状况。

夏季运行方式与冬季相反。图 5-48、图 5-49 中，高侧窗内侧玻璃应用扩散玻璃，使辐射热分布到更多表面，以免局部表面温度高，而过早向室内散热。

对于居住建筑，向阳面玻璃总面积与地板总面积之比可取 1/5～1/3。墙、地板以及其他热工构件的表面积应不少于 5 倍的向阳玻璃面积。

2) Direct gain system

Direct gain solar building lets solar radiation directly go into the room through transparent elements (e.g. windows) and indoor walls, floors, ground, furniture, as thermal mass, absorb and store heat in daytime, and supply heat at nighttime. For reducing heat loss, at night the windows should be covered with insulation curtain on cold side so as to avoid or greatly reduce condensation at the inside of the glass. If the surrounding gaps between the insulation curtain and the window are sealed airtight, the curtain may be positioned at warm side for convenient management. Fig. 5-48 and Fig. 5-49 indicate the operations in winter daytime and nighttime of the direct gain solar house.

In summer, the operation should be reversed from what is done in winter. Fig. 5-48 and Fig. 5-49 show the top window's inner glasses are diffusive glasses to distribute the radiation heat to more surfaces to avoid a small surface with high temperature to release heat indoor too early.

For dwellings, the ratio of glazed area facing the sun, to the total floor area may be between 1/5～1/3. The total surface area of walls, floors and other thermal mass should not be less than 5 times of sun facing glazing.

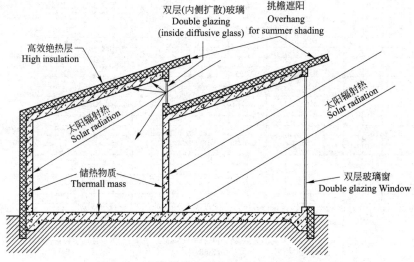

图 5-48 直接得热太阳房冬季白天工作状况

Fig. 5-48 Direct gain solar building's operation in winter daytime

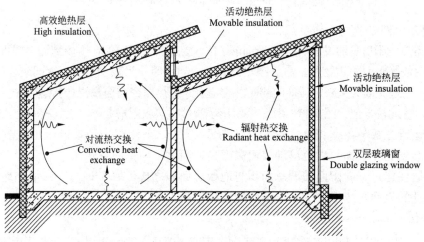

图 5-49 直接得热太阳房冬季夜间工作状况

Fig. 5-49 Direct gain solar building's operation in winter nighttime

3) 太阳能建筑选例

① 学校教室 英格兰北部华莱塞乔治街的二层学校建筑是一幢直接得热太阳房好例

5.2 建筑中应用太阳能
5.2 Application of Solar Energy in Buildings

子。图 5-50 为其剖面,向阳面全部双层玻璃,内侧为扩散玻璃。两玻间距 60cm,可装遮阳板,以遮挡夏季太阳辐射,同时作为走道供清扫。该建筑的贮热物质:钢筋混凝土屋面板 18cm 厚,楼板 23cm 厚,地板 25cm 厚,以及 23cm 厚的砖墙。屋面板和北墙外侧采用膨胀聚苯乙烯绝热层,13cm 厚(注:聚苯乙烯燃烧时有毒气现应禁止采用)。

记录资料表明,一年中该建筑获得的热量百分比:太阳能约 70%,灯光热约 22%,人体热约 8%。辅助供暖设备一直未用过。更令人感兴趣的是,该校所在地区冬季几乎无直射阳光,热能主要来自扩散辐射。

3) Selected examples of solar buildings

① School classroom. The St George's two-story Wallasey school building in Northern England is a good example of direct gain solar building. Fig. 5-50 shows its section. Sun facing façade is double glazed with inner layer of glass diffusible. The two panes are set 60cm apart to permit installation of shading boards for not only stopping summer solar radiation but also as walkway for cleaning. The thermal mass of the building is: 18cm thick reinforced concrete roof slab, 23cm thick floor slab, 25cm thick ground slab and 23cm thick brick wall. The roof slab and northern wall are covered with an exterior layer of expanded polystyrene insulation in 13cm thick. (Note: polystyrene in fire can release noxious gas and is now prohibited).

Recorded data indicated that in a year the building has received 70% of solar energy, 22% of lamp heat and 8% of human heat. Auxiliary heating device has never been operated. The more interesting point is that the region of the school almost has no direct sunshine in winters. The heat energy mainly comes from diffusive radiation.

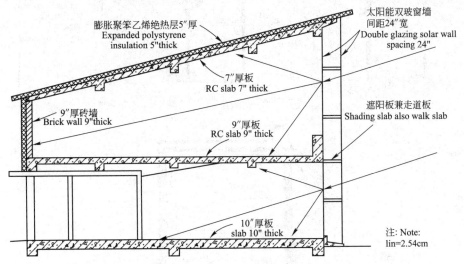

图 5-50 英格兰·乔治街学校太阳房剖面(建筑师:E. A. Morgan)
Fig. 5-50 Section through St George's School (Architect: E. A. Morgan).

② 住宅,图 5-51 为美国新墨西哥州一幢二层楼的毗连阳光间太阳房住宅。由剖面可见,该住宅日光间与二楼顶棚、北墙内侧空气循环通道及底层石块贮热床有通风道相连,以风扇强迫对流方式循环运行。冬季白天,日光间内被太阳热加热的空气沿上述循环通道流动,沿途将顶棚与北墙加热成低温辐射面,向室内供暖,剩余热量由石床贮存,以备夜

5 THE APPLICATION OF SOLAR ENERGY AND METHANE IN BUILDINGS

间供暖。分隔阳光间与主房间的南墙做成集热墙，冬季白天贮热，夜间供暖。日光间玻璃内侧设有半透明遮阳百叶，夏季白天关闭，减少进热，同时保持一定的自然采光；夜开增强散热。冬季则反之。

② Dwelling house. Fig. 5-51 shows a two-story solar dwelling house with an attached sunspace in New Mexico USA. From the section we can see a ventilation tunnel through the sunspace, second-floor's ceiling space, inside of the northern wall and the ground rock bed. Air is circulated by mechanical blower. In winter daytime, in the sunspace, the air heated by solar energy circulates through the above-mentioned channel to heat the ceiling and northern wall and make them low temperature radiative surface to heat the room. The remaining heat is stored by the rock bed for night heating. The southern wall between the sunspace and the main room is made as a thermal mass wall. In winter it stores heat during the day and releases the heat to heating the room at night. Against inside surface of sunspace glazing there is a semi-transparent shading shutter. In summer daytime it is closed to reduce heat gain while keep a certain day lighting and is opened at night to increase heat flow outwards. In winter it operates inversely.

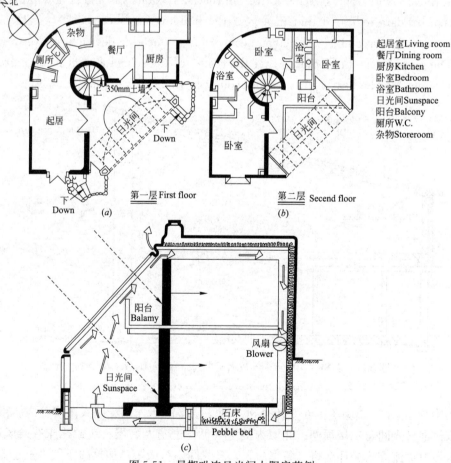

图 5-51　早期毗连日光间太阳房范例

Fig. 5-51　A good early example of solar house with attached sunspace

5.3 沼 气
5.3 Marsh Gas

③ 综合用能多层住宅,图 5-52 为一幢综合用能的多层住宅(作者为设计者之一)。

综合用能措施:ⓐ通过窗户直接得热与窗间墙间接得热相结合;ⓑ因西安地区太阳能不富足,故仍保留常规供暖作补充;ⓒ太阳能热水器装于阳台栏板上或窗间墙上;ⓓ利用厨房炊事弃热做暖墙;ⓔ采用毗连日光间(实验房间)。

通过上述措施。节能率达到 50%以上(未计入太阳热水的节能率)。

③ A multi-story dwelling using energy comprehensively. Fig. 5-52 shows a multi-story dwelling using energy comprehensively (the author is one of the designers).

Strategies of using energy comprehensively:

ⓐUse a combination of direct gain through windows and indirect gain through the walls between windows; ⓑsolar energy resource in Xi'an is poor so the normal heating is kept as an auxiliary system; ⓒSolar heaters installed at balcony rail panel or the walls between windows; ⓓ Build warm wall with cooking surplus heat; ⓔ Use attached sunspaces (in experimental rooms)

Through the above treatments the building achieved energy saving rate of more than 50% (not including solar hot water).

图 5-52 综合用能多层住宅
Fig. 5-52 A multi-story dwelling using energy comprehensively

5.3 沼气

5.3.1 沼气基本知识

沼气是一种以甲烷(CH_4)为主的混合气体。多种有机物在一定温湿度和厌氧条件下通过甲烷菌群的作用转化为沼气。在自然界,各种厌氧生态系统中都普遍存在微生物产生甲烷的现象。每年释放到大气中的甲烷多达几十亿吨,占大气中甲烷来源总量 90%。

沼气中一般含甲烷 60%,二氧化碳 35%,其余 5%为氮、一氧化碳、氢、硫化氢及微量氧。甲烷、一氧化碳和氢是可燃气体,通过燃烧产生热和光。

甲烷物理性质:

1) 无色、无味、无毒,但因有少量硫化氢(H_2S),略有蒜味或臭鸡蛋味。

2) 相对密度为 0.55(空气=1),比空气约轻一半。

3) 溶解度很低,20℃,一个大气压下体积溶解度约 3%。

4) 扩散速度比空气快 3 倍。熔点-182.5℃,沸点-161.5℃,着火点 537.2℃。

甲烷化学性质:

1) 完全燃烧时呈蓝色焰,最高温度可达 1400℃,放出大量热。燃烧化学反应式:

$$CH_4 + 2O_2 = CO_2 + 2H_2O + 881.3kJ(热量)$$

5 THE APPLICATION OF SOLAR ENERGY AND METHANE IN BUILDINGS

每 1m³ 甲烷在标准状态(0℃，1 个大气压)完全燃烧，可放出 35822.6kJ 热量。由于沼气中含甲烷 50%～70%，故每 1m³ 沼气完全燃烧可得 35822.6×(0.5～0.7)= 17911.3～25075·8kJ 热量，是同条件下煤气产热的 1.07～1.5 倍。

2) 空气按体积比混有 4%～5%甲烷时，遇明火即爆炸。

沼气池还能杀死多种病虫与病菌。

5.3 Marsh Gas

5.3.1 Basic knowledge of marsh gas

Marsh gas is a mixed gas but mainly of methane CH_4. Various organisms under a certain temperature, humidity and anaerobic condition, through the action of methane germs convert into marsh gas. In various natural anaerobic eco-systems there widely exist the phenomena of microbes generating methane. Every year, more than several billion tons of methane has been released into the air. This is 90% of the total methane source to the air.

Marsh gas general contains 60% methane and 35% CO_2 (carbon dioxide). The rest of 5% includes nitrogen(N), CO(carbon monoxide), hydrogen(H), H_2S(hydrogen sulphide) and a trace of oxygen(O_2). Methane, carbon monoxide and hydrogen are combustible gases and can generate heat and light through combustion.

Methane's physical characters:

1) Marsh gas has no color, none toxic and almost no smell except a little trait of garlic or rotten egg from hydrogen sulphide.

2) Its relative density is 0.55(Air=1), about 50% lighter than that of the air.

3) Low solubility-under 20℃ and one atmospheric pressure, its volumetric solubility is about 3%.

4) Its spreading speed is four times of the air. Its melting point (fusing point) is −182.5℃, boiling point −161.5℃, and ignition point(burning point) 537.2℃.

Methane's chemical characters:

1) In full burning it appears as blue flame with the highest temperature reaching 1400℃. In the meanwhile, large amount of heat is released. In burning its chemical reaction formula is:

$$CH_4 + 2O_2 = CO_2 + 2H_2O + 881.3kJ(heat)$$

Under the standard condition(0℃ and one atmospheric pressure) the methane in full burning can release heat at 35822.6kJ/m³. Marsh gas contains 50%～70% methane. So one cubic meter of marsh gas in full burning can supply heat at 35822.6×(0.5～0.7)= 17911.3～25075.8kJ which is 1.07～1.5 times that of gas under the same condition.

2) When air contains 4%～5% methane in volume rate, if contacts a burning fire, it will explode at once.

Methane-generating pit can kill various pests and germs.

5.3 沼 气
5.3 Marsh Gas

5.3.2 沼气池构造及工作原理

见图 5-53：原料和水经进料管"1"入发酵间"2"，经微生物分解产生沼气，上升到贮气间"3"，产气越多池内压力越大。当贮气间压力超过水压间"5"压力时，发酵液经导管"4"压入水压间"5"，产气越多，压入水压间（也叫出渣间）的发酵液也越多，并使其水位升高，而发酵间水位则下降，使两侧产生水位差压力，迫使沼气从导管"6"输出。若发酵间"2"产气不足，水压间"5"的水流回发酵间"2"，使水位升高，保持气体压力。若产气继续减少，两边水位将取平，供气停止。产气量与温度有关：50～55℃，每立方米池液产气 2～2.5m³；33～38℃，1～1.5m³；10～30℃，0.1～0.3m³。

5.3.2 Construction and operating principle of methane-generating pit

Refer to Fig. 5-53: Material and water go into the fermenting room "2" through input pipe "1". At "2" microbes decompose the material to generate marsh gas which rises up into the gas storage room "3". The more gas generated the higher pressure is in the room. In room "3" when the pressure is higher than that in the water pressure room "5", the higher pressure will press the fermented liquid into the room "5" via the pipe "4". The more gas generated the more fermented liquid pressed into the water pressure room (i.e. the dregs-output room) whose water level rises while the ferment room lowers its water level. That causes a pressure difference between the two sides. This pressure can force the marsh gas outward through the pipe "6". If generated gas is not enough, the water will flow back into room "2" to rise the water level and keep the gas pressure. If the gas continues to decrease, water at both sides will be at the same level and gas supply will be stopped. Gas output is relative to temperatures. When temperature reaches 50～55℃ it will generate gas at 2～2.5m³ per m³ fermented liquid, or 33～38℃, 1～1.5m³ and 10～30℃, 0.1～0.3m³.

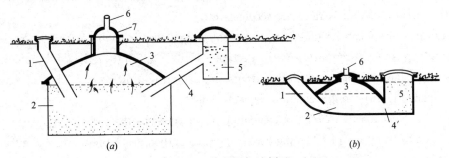

图 5-53 普通沼气池剖面

Fig. 5-53 Section of a common methane generating pit

1—进料管 Input pipe，2—发酵间 Fermenting room，3—贮气间 Gas storage room，4—导管 Pipe，4'—通口 An opening，5—水压间 Water pressure room，6—出气管 Gas out pipe，7—顶盖 Top cover

产气量与原料也有关，见表 5-3。

Gas output is relative to materials too, see table 5-3

5 太阳能与沼气在建筑中的应用
5 THE APPLICATION OF SOLAR ENERGY AND METHANE IN BUILDINGS

不同原料产沼气量　　　　　　　　　　　　　　　　表 5-3
Amount of methane generated by different materials　　Table 5-3

原料 Materials	每吨干原料产沼气(m³) One ton dry material generates methane(m³)	沼气中甲烷含量 CH₄ contained in methane(%)	原料 Materials	每吨干原料产沼气(m³) One ton dry material generates methane(m³)	沼气中甲烷含量 CH₄ contained in methane(%)
人粪 Human stool	240	50	麦秸 Wheat straw	432	59
厩肥 Barnyard Manure	260～280	50～60	玉米秸 Corn(maize)straw	250	53
牛粪 Cow dung	280	59	谷壳 Husk of rice	230	62
马粪 Horse dung	200～300	60	污泥 Sludge	640	50
猪粪 Pig dung	561	65	酒厂废水 Winery drainage	300～600	58
青草 Green grass	630	70	树叶 Tree leaves	210～294	58

　　沼气是较清洁的气体燃料，可用作炊事、照明、烘干、发电，或液化成甲醇取代石油。
　　沼气渣可作农肥、养殖物资：种蘑菇、养蚯蚓、养鱼、浸种、育秧等。
　　沼气渣由出渣间(水压间)开启顶盖取出。图 5-53(b)中"4"通口比(a)图"4"导管较便于清除发酵间积渣。发酵间需彻底清渣时，则需打开顶盖"7"进行。
　　图 5-54 是作者为某镇鸡、猪同楼及 200 余用户设计的太阳能沼气池，可延长供气期，获得较高产气量。主要原理：集热器"4"利用太阳能加热空气，由风机"6"送入石床"5"贮热，保持沼气池较长期的温暖，延长供气期，增加气量。鸡、猪楼设有自然空调通风道"1"；毗连种植温室"2"，使动、植物呼吸良性循环；屋面、墙面、散水种植有利于生态效益与环保。

　　Marsh gas is a relatively clean combustible gas. It can be used for cooking, lighting, drying and generating electricity. Also it can be liquified into methanol instead of petrol(gasoline).

　　Marsh gas dregs can be used as farm fertilizer, breed material to culture mushroom, earthworm, fishes and to soak seeds, raise rice seedlings, etc.

　　Marsh gas dregs can be taken out from the dregs-output room(water pressure room) by opening top cover. Comparing with the pipe "4" in Fig. 5-53(a) the "4'" opening in(b) is better for cleaning the accumulative dregs in the fermenting room. The top cover "7" must be opened if a thorough cleaning is required.

　　Fig. 5-54 shows a solar marsh gas generating pit, designed by the author, supplying more than 200 users in a town by use of a hen-pig house. It can lengthen gas supply term and increase output. Main principle: Collector "4" uses solar energy to heat the air. Blower "6" forces the warmed air into the rock bed "5" for heat storage so that the warmth in the pit can last a long term. Hence to lengthen gas supply days and increase output. The hen-pig building has natural air conditioning channel "1". It is attached to a plantable greenhouse which can make the animals and plants have a good circulation in breath. Landscaped roof, walls and apron area bring benefits to eco-system and environmental protection.

5.3 沼 气
5.3 Marsh Gas

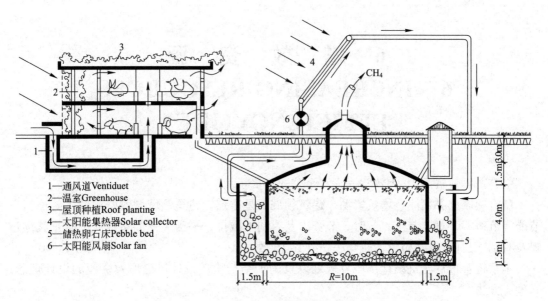

1—通风道 Ventiduet
2—温室 Greenhouse
3—屋顶种植 Roof planting
4—太阳能集热器 Solar collector
5—储热卵石床 Pebble bed
6—太阳能风扇 Solar fan

图 5-54 太阳能沼气池
Fig. 5-54 Solar methane generating pit

6 增效资源
6 INCREASING RESOURCES EFFICIENCY(IRE)

引言：增效资源 10 大效益——一个公式的揭示

所有领域都必须注意增效资源，建筑领域更是重点。建筑领域增效资源的主要项目是节能、节地、节材、节工、节时、节资以及有利于减排、平衡大气温室效应、防止气候过暖及有利于其他环保（防尘、减噪……）（共 10 效益）。

本章将着重从建筑设计、城乡规划以及围护结构构造设计讨论增效资源的设计理念、原理、方法。

本书所有各章均与上述 10 效益有关。

启发式

$$Q_{SHG}^{WHL}=\frac{t_i-t_o}{R_o}F \cdot Z \mp SRHG$$

Q_{WHL}——冬季失热；

Q_{SHG}——夏季得热；

t_i——室内气温；

t_o——室外气温；

R_o——围护结构总热阻；

F——建筑外表面积；

Z——热交换时间；

$SRHG$——太阳辐射进热。

Preface: IRE's 10 Benefits: Discover a Formula's True Meaning

All fields must pay attention to IRE. Architecture field is more important one. The main items of IRE in architecture field are: energy saving(ES), land saving(LS), material saving(MS), work saving(WS), time saving(TS) and finances saving(FS) and benefiting reducing discharges(RD), to balance AGE(BAGE) and preventing COW(PCOW) and other environmental protection effects(EP)(Altogether 10 benefits).

This chapter will focus on the ideas, principles and methods of building design, city-town planning and envelope detail design for IRE.

All chapters in this book are relevant to the above 10 benefits.

Enlightening formula

$$Q_{SHG}^{WHL}=\frac{t_i-t_o}{R_o}F \cdot Z \mp SRHG$$

Q_{WHL}——Winter heat loss;

Q_{SHG}——Summer heat gain;
t_i——indoor air temperature;
t_o——Outdoor air temperature;
R_o——The total thermal resistance of envelope;
F——External surface of the building;
Z——Heat exchange time;
$SRHG$——Solar radiation heat gain.

6.1 分析"F"

6.1.1 从"F"看建筑体形与增效资源

建筑平面基型有三：矩形、方形、圆形，如图 6-1(a)为三实例。

图 6-1(b)为 A、B、C 三幢建筑平面。它们建筑面积相同，均为 $10000m^2$，房高 10m，故体形体积（即外表面包容的体积）均为 $100000m^3$。但它们的体形不同，引起热交换面积不同，导致增效资源效益大异。设 F_a、F_b、F_c 分别为建筑 A、B、C 的外表面积，其值如下：

$$F_a = 2 \times (1000+10) \times 10 + 屋面\ Roofing\ 10000 = 30200m^2$$
$$F_b = 4 \times 100 \times 10 + Roofing\ 10000 = 14000m^2$$
$$F_c = 2\pi r \times 10 + Roofing\ 10000 = 10355m^2$$

6.1 Analyzing "F"

6.1.1 From "F" to Understand Building Shape and IRE

Building plans have three basic types: rectangle, square and round. Fig. 6-1(a) shows three examples.

Fig. 6-1(b) shows A, B and C three buildings. Their floor areas are the same of $10000m^2$ and height 10m. So their gross volumes (i.e. the volume contained by the external surface) all are $100000m^3$. Because their shapes are different. This leads to different areas of heat exchange that results in very different efficiency of IRE. Assume F_a, F_b, F_c are the external surface of building A, B and C respectively, their values are as follows:

$$F_a = 2 \times (1000+10) \times 10 + Roofing\ 10000 = 30200m^2$$
$$F_b = 4 \times 100 \times 10 + Roofing\ 10000 = 14000m^2$$
$$F_c = 2\pi r \times 10 + Roofing\ 10000 = 10355m^2$$

比较：

1) 用能：建筑 A 外表面积 F_a 最大，冬失热、夏得热最多，故其冬供热、夏供冷耗能最多，同时，其周边道路、管线及地下沟道最长。运行耗能也最多。简化表示比较：

耗能 $A>B>C$，节能 $C>B>A$；

2) 用地，建筑 A 占地（含绿地）最多，占地 $A>B>C$，节地 $C>B>A$；

3) 用材 $A>B>C$，节材 $C>B>A$；

6 增效资源
6 INCREASING RESOURCES EFFICIENCY (IRE)

4）用工 $A>B>C$，节工 $C>B>A$；

5）用时 $A>B>C$，节时 $C>B>A$；

6）用财 $A>B>C$，节财 $C>B>A$；

7）污染物（含 CO_2）排放 $A>B>C$；

8）对大气温室效应加剧作用 $A>B>C$；

9）对气候过暖加剧作用 $A>B>C$；

10）其他环保（防尘、减噪、空气质量……）负效应 $A>B>C$。

我们可用一个概念——建筑体形系数来评估上述 10 效益。

$$建筑体形系数 = \frac{建筑外表面积\ F}{体形体积\ V}$$

$$\text{Building A's BSC}ⓐ = \frac{F_a}{V} = \frac{30200}{100000} = 0.302$$

$$\text{Building B's BSC}ⓑ = \frac{F_b}{V} = \frac{14000}{100000} = 0.140$$

$$\text{Building C's BSC}ⓒ = \frac{F_c}{V} = \frac{10355}{100000} = 0.104$$

$$BSCⓐ > BSCⓑ > BSCⓒ$$

结论：体形系数越小越有利于增效资源的 10 效益。

Comparison:

1) Energy consumption: Building A's external surface Fa is the biggest, so its winter heat loss and summer heat gain are the most. Meanwhile its surrounding roads, pipelines and underground tunnels all are the longest, so their operating energy are the most.

To simplify the comparison:

Energy Consumption: $A>B>C$, Energy saving: $C>B>A$

2) Land-use, Building A occupies the most land, so land occupied (including landscape area), $A>B>C$, LS, $C>B>A$

3) Material-use, $A>B>C$, MS, $C>B>A$

4) Work-use, $A>B>C$, WS, $C>B>A$

5) Time-use, $A>B>C$, TS, $C>B>A$

6) Finances-use, $A>B>C$, FS, $C>B>A$

7) Pollutants discharge (incl. CO_2) $A>B>C$

8) The action of aggravating AGE, $A>B>C$

9) The action of aggravating COW. $A>B>C$

10) Other negative effects on the environment protection (dust stopping, noise reducing, air quality…) $A>B>C$

We can evaluate the above 10 benefits with a concept: Building Shape Coefficient (*BSC*).

$$BSC = \frac{Building\ external\ surface\ F}{Building\ shape\ volume\ V}$$

$$\text{Building A's BSC}ⓐ = \frac{F_a}{V} = \frac{30200}{100000} = 0.302$$

$$\text{Building B's BSC}ⓑ = \frac{F_b}{V} = \frac{14000}{100000} = 0.140$$

6.1 分析"F"
6.1 Analyzing "F"

$$\text{Building C's BSC}ⓒ = \frac{F_c}{V} = \frac{10355}{100000} = 0.104$$

$$\text{BSC}ⓐ > \text{BSC}ⓑ > \text{BSC}ⓒ$$

Conclusion: The less BSC the more beneficial to the 10 IRE

图 6-1

(a)矩形、方形、圆形建筑实例；①联合国办公楼；②商住楼；③福建土楼住宅群
(b)建筑面积相同，体形不同的三幢建筑 A、B、C 增效资源比较

(资料来源：靳亦冰)

Fig. 6-1

(a)Building examples of rectangle, square and round; ①UNOB; ②Apartment building; ③Fujian earth dwelling(Tulou) groups(b)The comparison of IRE between buildings A and B and C, that have same floor area but different shapes

(Source: Jin Yibing)

6 增效资源
6　INCREASING RESOURCES EFFICIENCY(IRE)

6.1.2　太阳能建筑体形系数

为得到更多太阳能，其体形系数应考虑方向性：

冬季

1）白天用的建筑东立面(Fe)可大些，通过窗户直接得热上午用；窗间墙贮热下午用。故东向体形系数 Fe/V 可大些。

2）夜间使用为主的房间如住宅，西立面(F_w)可大些，窗户直接得热下午用；窗间墙贮热晚上用。故西向体形系 F_w/V 可大些。

3）北半球南向体形系 F_s/V 和南半球北向体形系数 Fn/V 两者均可大些。

夏季，可采用近宅季青种植，浓叶遮阳，冬季叶落进热。还可用墙面绿化，水墙或充水墙等减少夏季进热。

随着科技的进步，用可转建筑获取最多太阳能将使梦成真(2006年张宜仁教授的建议。国外已有实验)

6.1.2　The Building Shape Coefficient(BSC) of solar buildings

In order to receive more solar energy, these buildings' BSC should relate to their orientations:

In winter

1) Building occupied mostly during the day should have larger east facade(F_e). Because windows allow direct gain for the space usage in the morning and the walls between windows(piers) store heat for afternoon use. So the east-orientated BSC F_e/V may be larger.

2) Building occupied mostly at night, such as dwelling, should have larger west facade (F_W). Because windows allow direct gain for afternoon use and the walls between windows stores heat for nighttime use. So the west-orientated BSC F_W/V may be larger.

3) Both south-orientated BSC F_s/V, in Northern Hemisphere and north-orientated BSC F_n/V in Southern Hemisphere may be larger.

In summer, we may use deciduous vegetation near house with thick leaves to shade the building. In winter, sun may shine into rooms because leaves are gone. We also can use landscaped wall, water wall or water-loaded wall to reduce summer heat gain.

Following the progress of science and technology, a dream may become true to use rotary building for maximizing the solar energy harvest. (Prof Zhang Yiren suggested it in 2006 year and there was an experiment in foreign countries recently).

6.1.3　从"F"看建筑结构、构造与增效资源

例：西安居住建筑，冬季保持室内气温20℃，在室外气温-6℃情况下，要求节能65%，若仍用黏土砖外墙，则需103cm厚，改用高效绝热材料——微孔泡沫塑料板，6cm就够了。在同样室内净面积下，砖外墙的建筑面积比塑料板的建筑面积四周要扩出97cm，如图6-2所示。按6.1.1同理分析可知，砖外墙建筑不利于增效资源，而高效绝热微孔泡沫塑料板的建筑有利于增效资源，获得10效益。我们可用一专业概念——建筑容积系数评估上述分析。

6.1 分析"F"
6.1 Analyzing "F"

$$建筑容积系数 = \frac{建筑外表面积 F}{建筑净容积 V_{net}}$$

V_{net}＝建筑体形体积V－建筑构件体积V_c

结论：建筑容积系数越小越显示该建筑结构和构造的先进性，越有利于增效资源的 10 效益。

6.1.3 From "F" to understand building structure, construction and IRE

Example：In Xi'an when outdoor temperature is -6℃ in winter, to keep a dwelling indoor temperature at 20℃, energy saving rate must reach 65％. If we still use clay brick external wall, its thickness must be 103cm. If we change to high performance insulation, say micro porous foamed plastic panel, 6cm thick is enough. Therefore, for same internal net area, the floor area of the brick building will be 97cm larger than the plastic panel building from all sides as Fig. 6-2 indicated. Similarly as analyzed in 6.1.1, we know that buildings with brick external wall do not benefit IRE. But buildings with high performance insulation micro porous foamed plastic panel do benefit IRE and have 10 benefits. We use a professional concept, Building Volume Coefficient (*BVC*), to evaluate the above analysis.

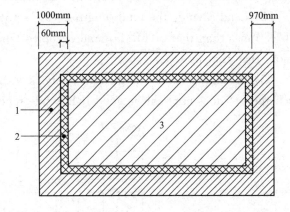

图 6-2 先进结构与构造有利于增效资源
Fig. 6-2 Advanced structure & construction benefit IRE
1—黏土砖墙 Clay brick wall 2—微孔泡沫塑料板 Micro porous foamed plastic panel 3—室内使用净面积 Internal net area

$$BVC = \frac{Building\ external\ surface\ F}{Building\ net\ volume\ V_{met}}$$

V_{net}＝Building shape volume V-Building component volume V_c

Conclusion：The less *BVC*, the more advanced structure and construction will be and the 10 benefits of IRE are more achievable.

6.1.4 从"F"看建筑布置与增效资源

例：某公司拟建 9 万 m² 纺织厂，设计者有两派：*A* 派主张分散布置，如图 6-3(*a*)，理由：通风好，采光好，噪声小，疏散快。

B 派主张集中布置，如图 6-3(*b*)，比较：

1) 节能：仅墙面面积 $F_{(a)} = 3F_{(b)}$，所以冬、夏耗能(*a*)＞(*b*)，若计入相关公用设施耗能，则(*a*)比(*b*)耗能更多。

2) 节地：建筑周围道路、管线及地下沟道总长(*a*)是(*b*)的 3 倍，所以，占地(含绿地)(*a*)比(*b*)多很多。

3) 总之，按 6.1.1 同理分析可知，(*a*)方案不利于增效资源，(*b*)方案有利于增效资源的 10 效益。

6 增效资源
6 INCREASING RESOURCES EFFICIENCY(IRE)

6.1.4 From "F" to understand building layout and IRE

Example: A company is going to build a 90,000m² textile mill. There are two groups of designers. Group A suggested a scattered layout scheme as shown in Fig. 6-3(a). Their idea is based on provision of good ventilation, good lighting, low noise and fast emergency escaping.

Group B suggested a compact layout scheme as shown in Fig. 6-3(b). In comparison:

1) Energy saving: Even if only counting the wall area, $F_{(a)}=3F_{(b)}$, there is difference of energy consumption between winter and summer, $(a)>(b)$. If adding the energy consumption of the relative public facilities, (a)'s energy consumption will be even more than that of (b).

2) Land saving: the total length of (a)'s roads, pipelines and underground tunnels is three times than that of (b). So land occupied (including landscaped area) by (a) is much more than that by (b).

3) In summary, according to the analysis with the same principles in 6.1.1 we see that scheme (a) does not achieve good IRE, while scheme (b) achieves the 10 benefits of IRE.

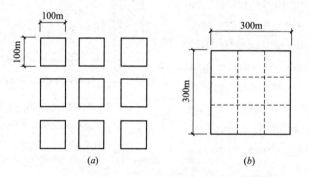

图 6-3 房屋分散布置与集中布置增效资源比较

Fig. 6-3 A comparison between scattered layout and compact layout of buildings in IRE

(a)每一车间 1 万 m²；(b)9 万 m² 大车间

(a)Each workshop 10000m²；(b)One big workshop 90000m²

(景观设计师金园园建议，中心块建开敞式庭院)

(Landscape designer Jin Yuanguan suggested to build an open courtyard in the center block)

4) 噪声：分散布置每车间 1000 台纺织机，集中布置的大车间 9000 台纺织机，哪个车间噪声影响人耳大呢？在此，有两条噪声规律起主控作用：一是距离衰减（空气分子对声能的吸收）；二是噪声的叠加与掩蔽规律。噪声叠加与掩蔽规律可简化如下：

$$\text{Assume two noises（设两噪声）} L_1 \geqslant L_2$$

$L_1-L_2=0\sim1(dB)$　　　Result: $L_1+3(dB)$

$L_1-L_2=2\sim3(dB)$　　　Result: $L_1+2(dB)$

$L_1-L_2=4\sim9(dB)$　　　Result: $L_1+1(dB)$

$L_1-L_2\geqslant10(dB)$　　　Result: $L_1+0(dB)$

例如：10 个噪声叠加，人耳的感受（单位：dB）见图 6-3′。

结论：影响工人的噪声只是其本身的纺织机噪声 L_1 以及与 L_1 声级差 $\Delta dB \leqslant 10dB$ 范围的噪声，故：(a)与(b)的噪声对工人影响没有什么差别。我们的调查，实验与理论分析

6.1 分析 "F"
6.1 Analyzing "F"

证实了上述结论。

4) Noise: Each workshop of the scattered scheme has 1,000 textile machines. The compact workshop has 9,000 textile machines. Which scheme has louder noise affecting human ear? Here, two principles of the sound mainly control the results. One is sound decay with distance (sound energy absorbed by air molecule), and the other is the principles of how sound piled and masked. It can be simplified as follows:

Assume two noises $L_1 \geqslant L_2$

$L_1 - L_2 = 0 \sim 1$ (dB)   Result: $L_1 + 3$ (dB)
$L_1 - L_2 = 2 \sim 3$ (dB)   Result: $L_1 + 2$ (dB)
$L_1 - L_2 = 4 \sim 9$ (dB)   Result: $L_1 + 1$ (dB)
$L_1 - L_2 \geqslant 10$ (dB)   Result: $L_1 + 0$ (dB)

Example: Human ears detect 10 noises piled (dB), see Fig. 6-3'

Conclusion: The noises affect the worker are only the noise from his own textile machine L_1 and the noise with sound level differences within 10 dB less than L_1. So there is hardly any difference of noise affecting workers between in scheme (a) and scheme (b). Our investigation, experiment and analysis have proved the above conclusion.

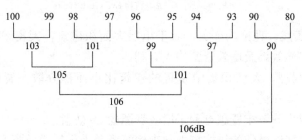

图 6-3′ 10 个噪声叠加，人耳的感受

Fig. 6-3′ Human cars detect the 10 noises piled

5) 通风换气：A 派认为，分散布置的车间，外墙距离短，容易得到穿堂风，故通风换气好，果真如此吗？

一般而言，穿堂风是可用的，但不总是有效的。热季室外气温高，穿堂风会使室内温度升高；冷季室外气温低，穿堂风会使室内温度降低，引起冬夏耗能更多。在沙尘天气，穿堂风更使室内热环境变坏，并降低产品质量。采用自然空调比穿堂风有利，大车间更便于采用地道风的自然空调，如图 6-4 所示。地下土层温度夏季比室外气温低，冬季比室外气温高，故室外空气经地道入车间可得冬升温、夏降温，并使其湿度得到调节。这就是自然空调的效果，地下风道还可作为紧急安全疏散通道。

5) Ventilation: Group A thinks that in scattered workshop the distance between the external walls is short. By opening the windows it can easily achieve cross ventilation. Is that true?

Generally cross ventilation is available but not always efficient. In hot seasons, outdoor air temperature is high. The cross ventilation will increase the indoor air temperature. In cold seasons, outdoor air temperature is low. Cross ventilation will decrease indoor air temperature. This will increase the energy consumption in winter and summer. In dusty weather, cross ventilation

6 INCREASING RESOURCES EFFICIENCY(IRE)

does make the indoor thermal environment very poor and reduce product quality. Utilization of natural air conditioning (NAC) is better than relaying on cross ventilation. In the large workshop, using ventilation in underground channels, as NAC is easier as Fig. 6-4 indicated. Underground soil temperature in hot seasons is lower than that of outdoor air. In cold seasons, it is higher than that of outdoor air. So after the outdoor air goes through the channel and entering the workshop, its temperature will be increased in winter and decreased in summer while the humidity is adjusted. That is the NAC effect. Underground ventilation channels also can be used as emergency safety escapes.

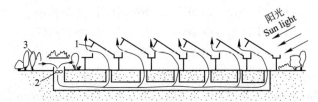

图 6-4　地道风自然空调系统

Fig. 6-4　A natural air conditioning system with underground ventilation channel

1—背阳锯齿形天窗 Saw tooth skylights at the side against the sun;

2—风机 Blower；3—常青植物 Evergreen plants

6）自然采光：实践、理论均证实，对于单层大面积建筑，天窗采光比分散的小面积建筑用墙窗采光，采光效果及增效资源均大有利。

7）环保：与前同理，大面积集中布置的建筑比小面积分散布置的建筑对环保有利得多。

结论：建筑物越少分散布置越有利于增效资源的 10 效益。

蜜蜂本能地将小窝以最佳的结构集中起来构筑成最优化的增效资源的大巢，我们早该学习它们了。

思考：一所大学所有院系能在一幢楼里办公，教学吗？

6) Natural lighting: Both practice and theory have proved that the efficiency and IRE of skylight lighting in a single storey large building, is much better than lighting from windows in the wall in scattered small buildings.

7) Environmental protection: from same reason as above, the compact layout of a large building achieves much better environment protection than the scattered layout of small building.

Conclusion: The less scattered layout of buildings is the better IRE's 10 benefits are.

With optimum structure, bees instinctively build many small hives into a big one with best IRE. We should have already learnt from them.

Thinking deeply: In one university, can all schools and departments work and teach in one building?

6.1.5　从"F"看城市形式与增效资源

我国西北兰州市，地处两山相夹的黄河南岸，形成了东西长 36km、南北最窄处仅 1km 的窄条形（图 6-5L），与西安、北京、莫斯科（图 6-5X、B、M）相比，显然，不利于增

6.2 分析"R"

6.2 Analyzing "R"

效资源的 10 效益。自 20 世纪 80 年代起,该市已在黄河北岸南山坡兴建靠山房和靠山掩土建筑,以改善该市长宽边之比。

设"P"为城市周边长,"A"为 P 包围的面积,那么,P/A(周面比)越小,则越有利于增效资源的 10 效益。

关于"F"的精歌:

(BSC)、(BVC)、(P/A)越小,以及分散布置房屋(SLB)越少,越有利于增效资源的 10 效益。

6.1.5 From "F" to understand city's form and IRE

Lanzhou city in the Northwest China is located at the south bank of Huanghe River and sandwiched between two mountains. It forms the city into a narrow shape with east-west length of 36km and minimum south-north width of 1km (Fig. 6-5L). Comparing with Beijing, Xi'an and Moscow (Fig. 6-5 B, X, M), it is obviously not good for the IRE's 10 benefits. Since 1980s, this city has been constructing hillside houses and hillside earth sheltered buildings at the southern side of the mountain along the north bank of the Huanghe River so as to improve the rate of the length to the width. If assume "P" as the perimeter of the city, "A" as the area surrounded by the P, then the less P/A ratio is, the better IRE's 10 benefits are.

"F" jingle

The less (BSC), (BVC), (P/A) and (SLB) the better IRE's 10 benefits.

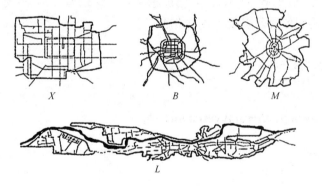

图 6-5 城市形式与增效资源

Fig. 6-5 City forms and IRE

L)兰州 Lanzhou; X)西安 Xi'an; B)北京 Beijing; M)莫斯科 Moscow

6.2 分析"R"

6.2.1 实有热阻 R_P

围护结构实有热阻(图 6-6),$R_p = R_i + R_w + R_e$ m²K/W,R_i:内表面热阻或称感热阻;R_e:外表面热阻或称放热阻。前苏联建筑科学研究院建议:

6 INCREASING RESOURCES EFFICIENCY(IRE)

$R_i=0.115m^2K/W$，即当室内空气高于内表面 1K 时，$1m^2$ 内表面从室内感受 1J 热量的时间是 0.115 秒；

$R_e=0.043m^2K/W$，即外表面高于室外空气 1K 时，$1m^2$ 面积向外释放 1J 热量的时间是 0.043s。

R_w——围护结构热阻，在图 6-6 中，为墙的热阻。一般而言，实体材料的热阻：

$$R=d/\lambda$$

d——材料厚度；

"λ"——材料热导率或称导热系数。其定义是：材料厚 1m，在一维传热条件下，两侧温差 1K，通过 $1m^2$ 面积，每秒由高温面传到低温面的热量焦耳数。"λ" 单位 W/(mK) 的推导：材料每秒(s)，由高温面传给低温面的热量焦耳(J)数(Q)与材料厚度(d)、面积(A)以及两侧温差(Δt)的关系可表达如下：

$$\frac{Q}{s}=\frac{\lambda A \Delta t}{d}$$

$$\lambda=\frac{Q}{s}\cdot\frac{d}{A\Delta t}$$

每秒焦耳数(Q/s)可用瓦特(W)表示，将 d、A、Δt 的单位：m、m^2、K 代入上式，可得 "λ" 的单位：

$$\frac{Wm}{m^2K}=W/(mK)$$

例：黏土砖墙 $\lambda=0.81W/(mK)$。其物理意义：当该墙厚 1m，两侧温差 1K，一维传热条件下，高温面传到低温面热量每秒 0.81J。

6.2 Analyzing "R"

6.2.1 Practical envelope thermal resistance R_p

The practical thermal resistance of an envelope (Fig. 6-6)

$R_p=R_i+R_w+R_e\ m^2K/W$

R_i: Inner surface resistance;

R_e: External surface resistance. It was suggested by the former CCCP Academy of Architectural Science that

$R_i=0.115m^2K/W$, i.e. when indoor air temperature is 1K higher than that of the inner surface, it takes 0.115 second for $1m^2$ inner surface to accept 1J heat from the room.

$R_e=0.043m^2K/W$, i.e. when the external surface temperature is 1K higher than that of the outdoor air, it takes 0.043 second for $1m^2$ external surface to release 1J heat.

R_w: Thermal resistance of the envelope. In Fig. 6-6, R_w is the wall's, resistance. In general, for a solid material, its resistance $R=d/\lambda$, d: the thickness of the material; "λ":

6.2 分析 "R"
6.2 Analyzing "R"

Thermal Conductivity of the material. It is defined as: in one second, how much heat, in joules, can be conducted from high temperature side to the low temperature side through a 1m² area of 1m thick material, under 1K temperature difference between both sides and under one-dimensional heat conduction. The inference of the unit of λ, W/(mK): a material in one second from high temperature side conducts how many joules heat (Q/s) to the low temperature side which is relating to the material's thickness(d), area(A) and the temperature difference(Δt) between the two sides, may show as follows:

$$\frac{Q}{s} = \frac{\lambda A \Delta t}{d}$$

$$\lambda = \frac{Q}{s} \cdot \frac{d}{A \Delta t}$$

The joules per second (Q/s) may be shown in watt (W), take the units of $d, A, \Delta t$: m, m², K in the above formula, we can get the unit of λ:

$$\frac{Wm}{m^2 K} = W/(mK)$$

Example: Clay brick wall $\lambda = 0.81 W/(mK)$. Its physical meaning is: When the wall is 1 meter thick, the two sides temperature difference is 1K, under the condition of one dimensional heat conduction, in every second, the high temperature side conducts 0.81J heat to the low temperature side.

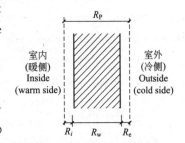

图 6-6 围护结构实有热阻
Fig. 6-6 Practical thermal resistance of envelope

影响"λ"的六要素：

1) 材料密度 ρ，一般说，"ρ"越大"λ"越大；建筑钢材 $\rho = 7850 kg/m^3$，$\lambda = 58.2 W/(mK)$；黏土实心砖墙 $\rho = 1800 kg/m^3$，$\lambda = 0.81 W/(mK)$。

2) 材料湿度：材料含有空气的微孔被水浸占，湿度就增加，空气 $\lambda = 0.029$，水的 $\lambda = 0.58$，是空气的 20 倍，故材料湿度增加，"λ"也增大。

3) 材料内微孔状态：图 6-7 示出材料三种内部微孔状态（显微镜下放大 100 倍），以 (a)、(b) 为例：(a) 内微孔密闭互不相通，(b) 内部分微孔相通，结果 $\rho_a > \rho_b$，而 $\lambda_a < \lambda_b$，因为，微孔相通造成对流传热加大。但 (b) 材料吸声较好。因为微孔相通，声能摩擦转热多。

4) 传热方向：图 6-8 示出的是：以木材为例，顺年轮纹 λ_{11} > 垂直于年轮纹的 $\lambda_\perp$。因为顺纹内水较多，传热快。

5) 材料分子、原子、电子参与传热的活跃性：

例：钢材密度 $\rho_s = 7850 kg/m^3$，导热系数 $\lambda_s = 58.2 W/(mK)$。铝材密度 $\rho_a = 2600 kg/m^3$，是 ρ_s 的 1/3，铝材的导热系数 $\lambda_a = 190 W/(mK)$，是 λ_s 的 3 倍多。这儿并不是密度 ρ 越大，"λ"也越大。原因是铝材分子、原子、电子参与传热的活跃性强，正如铜比铁导电性强，就是因为铜材电子参与导电的活跃性比铁电子强（铜的电子碰撞传电快）。

6) 时间：杜甫诗"布衾多年冷似铁"就说明时间影响"λ"值的例子。

夹心墙用防水预制绝热板填充比散粒填充好，可保持较长期"λ"原值不变。

6 增效资源
6 INCREASING RESOURCES EFFICIENCY(IRE)

Six main factors affecting "λ" are:

1) Material density ρ: In general, the bigger "ρ" is, the bigger "λ" is. For example, building steel's $\rho = 7850 kg/m^3$, $\lambda = 58.2 W/(mK)$, clay brick solid wall's $\rho = 1800 kg/m^3$, $\lambda = 0.81 W/(mK)$.

2) Material humidity: When water occupies the material's micro porous part, which was used to be filled with air, the material humidity will be increased and its "λ" will be increased too. Because the water's $\lambda = 0.58$ is twenty times the λ of the air (air $\lambda = 0.029$).

3) Micro porous state inside the material: Fig. 6-7 shows three micro porous states inside materials (magnified 100 times under a microscope). Taking (a) and (b) as an example. Inside (a), the micro voids closed from each other without any interpenetration. Inside (b), some micro voids are interpenetrated. It resulted $\rho_a > \rho_b$, but $\lambda_a < \lambda_b$. Because the interpenetrated micro voids increase the convective heat exchanges. But material (b) sound absorption is better because the interpenetrated voids can convert more sound energy into heat by friction.

4) Direction of heat conduction: Fig. 6-8 indicates that: if taking a wood as an example, the λ_{11} along the annual rings is larger than the $\lambda_\perp$ crossing the annual rings. Because there is more water between the annual rings, the conduction of heat is faster.

5) Active level of heat conductivity of material molecule, atoms and electrons:

Example: Steel's density $\rho_s = 7850 kg/m^3$, its conductivity $\lambda_s = 58.2 W/(mK)$. Aluminum's density $\rho_a = 2600 kg/m^3$, which is one third of the ρ_s, aluminum's conductivity $\lambda_a = 190 W/(mK)$, which is three times more than λ_s. Here, it is not true that the larger ρ is the larger λ will be. It is because that the aluminum molecule, atoms and electrons have stronger heat conducting activity. It is the same as that the cuprum (copper)'s electrical conductivity is stronger than that of iron. Because the copper electrons' electric activity is stronger than that of iron (copper electrons conduct electricity faster by impact).

6) Time: a famous poem by Dupu reads as: "A cotton-cloth quilt became cold like iron after many years". It's just an example explaining time affects "λ" value.

Sandwich wall filled with prefabricated waterproofing insulation board is better than that filled with loose insulation because it can keep the original value of "λ" for a long time.

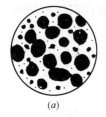

(a) (b) (c)

图 6-7 材料内微孔状态(放大 100 倍)
Fig. 6-7 Micro porous states in materials (Magnified 100 times)

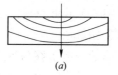
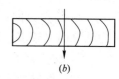
(a) (b)

图 6-8 "λ" 与传热方向关系
Fig. 6-8 "λ" and heat flow direction
$\lambda_a < \lambda_b$ i.e. $\lambda_\perp < \lambda_{11}$

表 6-1 为一些常见材料 "λ" 值。
Table 6-1 shows the λ values of some common materials

6.2 分析"R"
6.2 Analyzing "R"

常见材料"λ"值 表 6-1
λ of some common materials Table 6-1

材料 Materials	ρ kg/m³	λ W/(mK)	材料 Materials	ρ kg/m³	λ W/(mK)
钢材 Steel	7850	58.2	石棉水泥板 Asbestos cement board	1800	0.52
铝材 Aluminum	2600	190.0	石棉水泥绝热板 Asbestos cement insulation board	500	0.16
钢筋混凝土 Reinforced concrete	2500	1.74	石膏板 Gypsum board	1050	0.33
黏土砖砌体 Clay brick masonry	1800	0.81	水泥刨花板 Cement particle board	1000	0.34
水泥膨胀珍珠岩 Cement expanded perlite	800	0.26	稻草板 Straw board	700	0.19
聚乙烯泡沫塑料 Polyethylene foamed plastics	100	0.047	木屑板 Chip hoard	300	0.105
胶合板 Plywood	600	0.17	木屑 Wood chips(dust)	250	0.093
软木板 Corkboard	300 150	0.093 0.058	稻壳 Rice husk(hull)	120	0.06

- 确定"λ"值最好的办法是实验，不要轻信广告。
- The best method for determining "λ" is by experiment. Do not readily believe advertisings.

6.2.2 节能热阻 R_{ES}

我国自 2007 年起，要求居住类建筑冬季以 20 世纪 80 年代耗能为基准节能 65%，为此我们求出一节能热阻一般式如下：

$$R_{o\cdot min\cdot ES} = \frac{100}{100-n} R_{o\cdot min\cdot N} \quad m^2 K/W \tag{6-1}$$

式中：
$R_{o\cdot min\cdot ES}$：节能最小总热阻 m²K/W；
n：节能率中的数字，例如要求节能 65%，即 $n=65$。
$R_{o\cdot min\cdot N}$ 20 世纪 80 年代我国设计围护结构的低限热阻

$$R_{o\cdot min\cdot N} = \frac{t_i - t_o}{[\Delta t]} R_i \quad m^2 K/W \tag{6-2}$$

t_i：室内气温℃；
t_o：室外气温℃；
R_i：内表面热阻 $R_i = 0.115 m^2 K/W$；
$[\Delta t]$：室内气温与围护结构内表面温度容许差值。

6.2.2 Energy saving thermal resistance R_{ES}

Since 2007, China demands that in winter, residential buildings should save 65% of energy based on the energy consumption in 1980s. For this we have worked out a general formula of thermal resistance of energy saving as follows:

$$R_{o \cdot min \cdot ES} = \frac{100}{100-n} R_{o \cdot min \cdot N} \quad m^2 K/W \tag{6-1}$$

In the formula:

$R_{o \cdot min \cdot ES}$: The minimum total energy saving thermal resistance $m^2 K/W$;

n: the number of the energy saving percentage, such as ES 65%, i.e. $n=65$.

$R_{o \cdot min \cdot N}$ The minimum thermal resistance for designing building envelope in 1980s in China

$$R_{o \cdot min \cdot N} = \frac{t_i - t_o}{[\Delta t]} R_i \quad m^2 K/W \tag{6-2}$$

t_i: Indoor air temperature ℃;

t_o: Outdoor air temperature ℃;

R_i: Inner surface thermal resistance $R_i = 0.115 m^2 K/W$;

$[\Delta t]$: Allowable temperature difference between indoor air and the inner surfaces of envelope.

[Δt] 例 表 6-2
Examples: [Δt] Table 6-2

建筑类型 Types of buildings	外墙 Exterior wall, ℃ or K	屋顶 Roof ℃ or K
居住建筑、医院、托幼 Residential building, hospital, nursery/kindergarten	6.0	4.0
办公、学校、门诊 Office building, school, clinic building	6.0	4.5
公共建筑 Public building	7.0	5.5

[例1]: 求西安地区某居住建筑(砖混结构)$R_{o \cdot min \cdot N}$, 并求外墙厚 d(m)。设 $t_i = 20℃$, $t_e = -6℃$ 采用黏土砖墙，剖面如图 6-9。

[解]

$$R_{o \cdot min \cdot N} = \frac{t_i - t_o}{[\Delta t]} R_i = \frac{20-(-6)}{6} \times 0.115$$
$$= 0.498 \doteq 0.50 \quad m^2 K/W \tag{6-3}$$

该墙总热阻:

$$R_o = R_i + R_1 + R_2 + R_3 + R_e$$
$$= 0.115 + \frac{0.02}{0.81} + \frac{d}{0.81} + \frac{0.02}{0.93} + 0.043$$
$$= \frac{d}{0.81} + 0.204$$

6.2 分析 "R"
6.2 Analyzing "R"

令 $R_o = R_{o \cdot min \cdot N}$，即 $\dfrac{d}{0.81} + 0.204 = 0.50$

$$d = (0.50 - 0.204) \times 0.81 = 240 \text{mm}$$

取 240mm，即 24 墙。当今西安地区绝大多数建筑外墙仍是 24 墙，就是那时建筑热工要求定的。现在这种墙早已不满足节能要求了。

[例 2]：按节能 65% 设计西安地区居住建筑外墙。

[解]

$$R_{o \cdot min \cdot ES} = \dfrac{100}{100-n} R_{o \cdot min \cdot N}$$

$$= \dfrac{100}{100-65} \times 0.5 = 1.43 \quad \text{m}^2\text{K/W}$$

如图 6-9，$R_o = R_i + R_1 + R_2 + R_3 + R_e$

$$= 0.115 + \dfrac{0.02}{0.81} + \dfrac{d}{0.81} + \dfrac{0.02}{0.93} + 0.043$$

$$= \dfrac{d}{0.81} + 0.204$$

令 $R_o = R_{o \cdot min \cdot ES}$　i.e. $\dfrac{d}{0.81} + 0.204 = 1.43$

$$d = (1.43 - 0.204) \times 0.81$$

$$= 0.993 \text{m} \doteq 1\text{m}$$

谁也不会用 1m 厚的墙去满足节能 65%，因为：

1) 太重 (1246kg/m²) 不能用；
2) 结构面积太大不能用，怎么办？

复合墙应时而生。复合墙乃优势互补墙

[Example No. 1]: To decide the $R_{o \cdot min \cdot N}$, and design the external wall thickness d (m) of a concrete-brick construction of a residential building, in Xi'an, we use clay brick wall. Its section is shown in Fig. 6-9.

[Solution]

$$R_{o \cdot min \cdot N} = \dfrac{t_i - t_o}{[\Delta t]} R_i = \dfrac{20 - (-6)}{6} \times 0.115$$

$$= 0.498 \doteq 0.50 \quad \text{m}^2\text{K/W} \tag{6-3}$$

The total thermal resistance:

$$R_o = R_i + R_1 + R_2 + R_3 + R_e$$

$$= 0.115 + \dfrac{0.02}{0.81} + \dfrac{d}{0.81} + \dfrac{0.02}{0.93} + 0.043$$

$$= \dfrac{d}{0.81} + 0.204$$

Set $R_o = R_{o \cdot min \cdot N}$, i.e. $\dfrac{d}{0.81} + 0.204 = 0.50$

$$d = (0.50 - 0.204) \times 0.81 = 240 \text{mm}$$

If name the wall with thickness of 240mm as 24 wall. Today in Xi'an region, most build-

6 INCREASING RESOURCES EFFICIENCY (IRE)

ings' external walls are still 24 walls with clay bricks. It was decided by the thermal requirements then. Now, the 24 wall is no longer meeting the demand of energy saving.

[Example No. 2]: To design an external wall of a residential building in Xi'an region, it demands 65% of energy saving.

[Solution]

$$R_{o \cdot min \cdot ES} = \frac{100}{100-n} R_{o \cdot min \cdot N}$$

$$= \frac{100}{100-65} \times 0.5 = 1.43 \quad m^2 K/W$$

As in Fig. 6-9, $R_o = R_i + R_1 + R_2 + R_3 + R_e$

$$= 0.115 + \frac{0.02}{0.81} + \frac{d}{0.81} + \frac{0.02}{0.93} + 0.043$$

$$= \frac{d}{0.81} + 0.204$$

Set $R_o = R_{o \cdot min \cdot ES}$ i.e. $\frac{d}{0.81} + 0.204 = 1.43$

$d = (1.43 - 0.204) \times 0.81$

$= 0.993m \doteq 1m$

No one will use 1m thick wall to satisfy 65% of energy saving target because:

1) Too heavy (1246kg/m²) to use;

2) Too big area of structure to use. What to do?

Composite walls (CW) emerge on time. Composite walls are the walls with components' merits complementing each other.

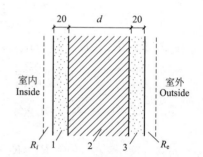

图 6-9 本例题外墙剖面
$R_i = 0.115$, $R_e = 0.043$, 1: 石灰砂浆抹灰, 20, $\lambda = 0.81$; 2: 黏土砖墙, $\lambda = 0.81$, 3: 水泥砂浆抹灰, 20, $\lambda = 0.93$

Fig. 6-9 The exterior wall section of the example, $R_i = 0.115$, $R_e = 0.043$, 1: Lime mortar, 20, $\lambda = 0.81$; 2: Clay brick wall, $\lambda = 0.81$, 3: Cement mortar, 20, $\lambda = 0.93$

重型复合墙

图 6-10 展示了三种重型复合墙：(a) 黏土砖夹泡沫塑料夹心墙；(b) 砖啮接夹心墙；(c) 国外几例钢筋网片连接的复合墙，钢筋直径 4～6mm。砖咬接或钢筋网片连接均属冷 (热) 桥连接，因为该处热阻比复合部分小，冬失热多，夏进热多。为改进可用穿孔塑片带连接 (图 6-11(b)、(c)～图 6-14)，塑片厚 0.6～0.8mm，只要塑片热导率小于或等于绝热层的热导率就是无冷 (热) 桥连接 (国外已有实例)。塑片穿孔是便于砂浆将上下层砖连接起来。更进一步的改进就是无砂浆连接，将复合墙预制成砌块，现场用高绝热粘合剂粘合或制成大型钢筋混凝土壳夹心墙板或压型钢 (铝) 板夹心墙板。

重型复合墙可承重盖6层及6层以下房屋，并可满足节能要求。

Heavy composite walls

Fig. 6-10 shows three types of heavy composite walls: (a) clay brick sandwich wall filled with foamed plastics; (b) brick inter-jointed sandwich wall; (c) sandwich walls reinforced by steel mesh in foreign examples with steel bar of 4～6mm diameter. Joints with brick ties or reinforcing mesh are cold (heat) bridge joints. Because the thermal resistance here is less than that of the composite part. The thermal bridge can increase heat loss in winter and heat gain in summer. We may use perforated plastic tie (Fig. 6-11(b), (c), 6-12, 13 and 14) to improve this. The thickness of the tie is 0.6～0.8mm. If the conductivity of the tie is

6.2 分析 "R"
6.2 Analyzing "R"

less than or equal to that of insulation, the joints are not cold (heat) bridge joints anymore. (There are examples in foreign countries). Plastic tie has holes to allow mortal to joint the bricks up and down. A further improvement is to have no mortal joint. Composite wall can be prefabricated into blocks and adhere together on site with high adiabatic adhesive. Alternatively, prefabricate the composite wall into big reinforced concrete shell sandwich panels or pressed steel (aluminum) sandwich panels.

Heavy composite walls can be used as load bearing walls for buildings egualling or under 6-story while satisfying the energy saving requirements.

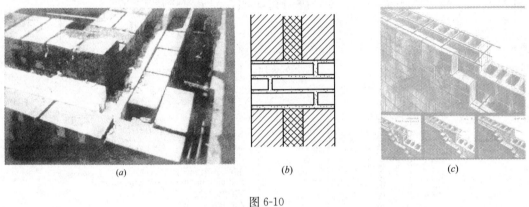

图 6-10

Fig. 6-10

(a)黏土砖夹泡沫塑料复合墙(张家璋教授提供);(b)砖啃合(咬接)夹心墙;(c)钢筋网片连接夹心墙

(a)Clay brick sandwich wall with foamed plastics(Source:Prof Zhang Jiazhang);(b)Brick bridle (joggle) sandwich wall;

(c)Sandwich walls with reinforced mesh

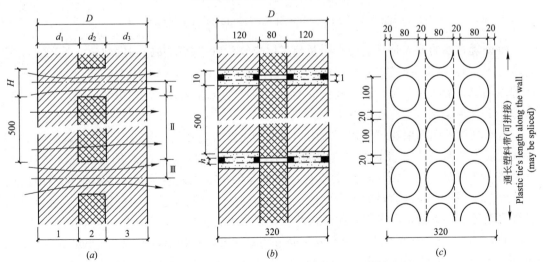

图 6-11 有冷(热)桥和无冷(热)桥的夹心墙

Fig. 6-11 Sandwich walls with and without thermal bridge joints

(a)砖咬接有冷(热)桥效应;(b)、(c)穿孔塑料片连接无冷(热)桥效应

(a)Having thermal bridge effect;(b)、(c)Perforated plastic tie joints without thermal bridge effect

6 增效资源
6 INCREASING RESOURCES EFFICIENCY(IRE)

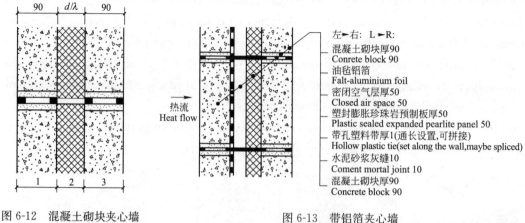

图 6-12 混凝土砌块夹心墙
Fig. 6-12 Concrete block sandwich wall

图 6-13 带铝箔夹心墙
Fig. 6-13 Sandwich wall with aluminum foil

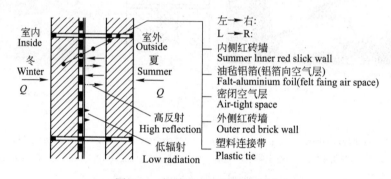

图 6-14 铝箔——空气层复合墙
Fig. 6-14 Composite wall with aluminum foil-air gap

重型复合墙的缺点是不适合高层及超高层建筑。对于高层和超高层建筑宜采用轻型复合墙。图 6-15 为某高层住宅轻型复合外墙正在安装，由外→内：压型钢板（厚 0.6mm）、矿棉绝热层、密闭空气层、石膏板、糊墙布。轻型复合墙每 m² 只有几十千克，而重型复合墙每平方米是几百千克甚至大于 1t。

近来高层及超高层建筑较盛行采用钢筋混凝土框剪墙外贴高绝热层，可以满足以 1980 年代耗能为基准节能 65% 的要求。西安一例，该高绝热层为 6cm 厚的微孔泡沫塑料板，外贴防水纤维布。有一种结构正受重视：防火涂层钢框架垂直承重，压型钢板木地面楼层，外包墙用防水高绝热轻型复合墙板。这种结构从一层到超高层均可用，最大优点是钢材可再循环，再更新，再应用，无建筑垃圾，是重大灾害中危害最小的一种轻结构。

The heavy composite wall is not suitable to high·rise and super high-rise buildings, which are better to use light composite walls. Fig. 6-15 shows a high·rise dwelling's light composite exterior wall under construction. From outside to inside: the wall consists of pressed steel sheet (t: 0.6mm), mineral wool insulation, closed air space, gypsum board and wall paper. Light composite wall only has several tens kg/m². In comparison, heavy composite wall has several hundreds kg/m² to somewhat more than one ton/m².

6.2 分析 "R"
6.2 Analyzing "R"

In recent years the reinforced concrete frame and shear wall structure with attached high insulation externally has been popularly used for high-rise and super high-rise buildings. It can satisfy the 65% of energy saving target based on the energy consumption of 1980s. Taking an example in Xi'an, the high insulation is a micro void foamed plastic board of 6cm thick with a waterproof fiber cloth adhered to its external surface.

A structure has been receiving more and more attention recently. It consists of steel frames with fireproof coating as vertical load bearing members, floor structure made of pressed steel sheet and timber flooring and light composite wall panel with waterproofing and high performance insulation as the building external envelope. It is suitable to buildings from single storey to super high-rise ones. Its best feature is that the steel can be recycled, renewed and reused without construction waste. Its light construction has minimum impact in serious calamities.

图 6-15 轻型复合墙正在安装
Fig. 6-15 A light composite wall under installation

6.2.3 复合墙剖面设计

复合墙剖面设计主要是从建筑热工考虑确定高绝热层。现以图 6-12 为例，绝热层两侧混凝土砌块承重各厚 10cm，$\rho=2100 kg/m^3$，$\lambda=1.28 W/(mK)$，试确定其绝热层。要求节能 65%，地点：西安。

[解]

$$R_{o \cdot min \cdot ES} = \frac{100}{100-n} R_{0 \cdot min \cdot N} \quad m^2 K/W$$

$$R_{o \cdot min \cdot N} = 0.50 (见 [例 1])$$

We have: $R_{0 \cdot min \cdot ES} = \frac{100}{100-65} \times 0.50 = 1.43 \quad m^2 K/W$

此处墙的总热阻是 R。

$$R_o = R_i + R_1 + R_2 + R_3 + R_e \quad R_2 = d/\lambda$$

令

$$R_o = R_{o \cdot min \cdot ES}$$

i.e. $0.115 + \frac{0.10}{1.28} + \frac{d}{\lambda} + \frac{0.10}{1.28} + 0.043 = 1.43$

$$\frac{d}{\lambda} = 1.43 - 0.314 = 1.116$$

(1) Assume 设定 "d" to select 选 "λ" 设 $d=10cm$

于是，

$$\lambda = \frac{0.10}{1.116} = 0.090 \quad W/(mK)$$

由表 6-1：聚乙烯泡沫塑料，$\rho=100$，$\lambda=0.047$；稻壳 Rice husk $\rho=120$，$\lambda=0.06$ 两者均能满足要求。

6 增 效 资 源
6 INCREASING RESOURCES EFFICIENCY(IRE)

(2) 如果只能买到一种绝热材料，例如聚乙烯泡沫塑料，$\rho=100$，$\lambda=0.047$. 就可算出：$d=1.116\times"\lambda"=1.116\times0.047=0.052$m i.e. $d=5.2$cm

6.2.3 Section design of composite wall

The main criterion of designing section of composite wall is to determine high performance insulation in accordance with thermal principles. Fig. 6-12 shows an example. On both sides of the insulation there is a layer of 10 cm thick load bearing concrete blocks. $\rho=2100$kg/m^3, $\lambda=1.28$W/(mK). The question is how to decide the required insulation in order to save 65% of energy in Xi'an.

[Solution]

$$R_{o\cdot min\cdot ES}=\frac{100}{100-n}R_{o\cdot min\cdot N} \quad m^2K/W$$

$R_{o\cdot min\cdot N}=0.50$ (see N0. 1 example)

We have: $R_{o\cdot min\cdot ES}=\frac{100}{100-65}\times0.50=1.43 \quad m^2K/W$

Here the total thermal resistance of the wall is R_o

$$R_o=R_i+R_1+R_2+R_3+R_e \quad R_2=d/\lambda$$

Set $R_o=R_{o\cdot min\cdot ES}$

i.e. $0.115+\frac{0.10}{1.28}+\frac{d}{\lambda}+\frac{0.10}{1.28}+0.043=1.43$

$$\frac{d}{\lambda}=1.43-0.314=1.116$$

(1) Assume "d" to select "λ"   If $d=10$cm

Then, $\lambda=\frac{0.10}{1.116}=0.090 \quad W/(mK)$

From Table 6-1: Polyethylene foamed plastics, $\rho=100$, $\lambda=0.047$; Rice husk $\rho=120$, $\lambda=0.06$ can both satisfy the need。

(2) If we can only buy an insulation material such as polyethylene foamed plastics, $\rho=100$, $\lambda=0.047$. We'll have $d=1.116\times"\lambda"=1.116\times0.047=0.052$m i.e. $d=5.2$cm

6.2.4 复合墙增热阻不增重或微增重

1) 在复合墙内加一层密闭空气层(3～5cm厚)就可增加热阻0.18m^{2}K/W，相当于23cm厚混凝土的热阻，而重量几乎没有增加。

2) 图6-15为在密闭空气层冬季暖侧贴一层油毡铝箔的铝箔——空气层复合墙，铝箔面向空气层，油毡可增强铝箔机械强度，并阻止水蒸气渗透。

铝箔热工特性(图6-14)是高反射低辐射。冬季：热流向外，铝箔辐射散热只有红砖的1/3，减少热损失2/3。夏季：铝箔高反射，进热只是红砖的1/3，减少进热2/3。该复合墙附加了铝箔与密闭空气层相当于增加了37墙的热阻。

使用铝箔要诀：一空、二暖、三净。即：要有空气层，要靠冬季暖侧(防止水蒸气进入，油毡水蒸气阻力高)；要干净(在生产、运输、安装全过程必须保持干净)。

6.2 分析 "R"

6.2 Analyzing "R"

6.2.4 To increase CW's thermal resistance without adding any weight or slightly increasing weight

1) In a composite wall, we can increase 0.18m²K/W of thermal resistance by adding a closed air gap (3~5cm thickness). It is equal to the thermal resistance of a 23 cm thick concrete wall without adding much weight.

2) Fig. 6-15 is a composite wall with sealed air gap and a layer of felt-aluminium foil adhered to the warm side in winter. The felt can increase aluminium foil's mechanic strength and stop vapor penetration.

Aluminium foil's thermal character(Fig. 6-14)is high reflection and low radiation. In winter, heat flows to outside. Aluminium foil's radiating heat is only 1/3 of that of red brick therefore it decreases 2/3 of heat loss. In summer, due to the high reflection of aluminum foil, the heat gain is only 1/3 of that of red brick. Therefore it decreases 2/3 of heat gain. This composite wall with additional aluminium foil and a sealed air gap is corresponding to an increase of thermal resistance of a 37cm thick clay brick wall.

The key points of applying aluminium foil are: one air gap, against warm side in winter (to stop vapor penetration because of high vapor resistance of felt) and clean (must be kept clean in the process of production, transportation and installation).

6.2.5 旧墙体改造

改造广大城乡旧墙使满足节能要求是一项简单而又艰巨的工程。说简单，是从技术上只要加一层绝热层就行。说困难，是全国仅旧住宅就有400多亿 m²，完成改造何其艰巨也?!

加绝热层有两法：在外墙上外包或内贴。

哪个好呢？先了解温度梯度的概念。图6-16(a)为绝热差的混凝土块，(b)为绝热好的塑料板，在同样热阻值($R_1 = R_2$)时，必然 $d_1 > d_2$。由于 $\Delta t_1 = \Delta t_2 \therefore (\text{tg}a_1 = \Delta t_1/d_1) < (\text{tg}a_2 = \Delta t_2/d_2)$，单位是 K/m，即单位距离温度降落的度数。这就叫温度梯度。

记住口诀："绝热差，温降缓。绝热好，温降陡"，就会容易作出图6-16(c)、(d)所示温度梯度曲线，(c)为外包(冬冷侧)绝热层，(d)为内贴(暖侧)绝热层。两图中，混凝土同厚，绝热层同厚。设室内气温20℃，室外气温－10℃，两图中混凝土取同一缓斜率，绝热层取同一陡斜率，即可作出两图温降曲线，并与0℃水平线交于两图中各自的"m"点。

分析：一个重要概念：材料层数相同，总热阻相等，布置不同，温度分布大异。图(c)外包(绝热层在冷侧)有下列优点：

1) 负温区小，内部冷凝危险小；
2) 平均温度高，维持室温稳定性较好。

适用于连续使用的房间，如住宅、敬老院……但外包施工较困难。(d)图，绝热层内贴(在暖侧)施工较方便。适用于间歇使用房间，如学校、商店、影剧院等。绝热层在暖侧，墙体吸热少，室内升温快。

6.2.5 To remodel old walls

To remodel the old walls in vast cities and villages to satisfy the energy saving demand

6 INCREASING RESOURCES EFFICIENCY(IRE)

is a simple but arduous project. Saying it is simple because technically it only needs to attach a layer of insulation. Saying it's arduous because there are more than 40 billion m² just old dwellings in China. How tremendous this task is?!

There are two ways to attach insulation: on outside or inside of external wall.

Which way is better? First we need to understand the concept of temperature gradient. Fig. 6-16(a) is a concrete block with poor insulation. (b) is a high performance insulation plastic board. Under the same resistance value ($R_1 = R_2$), it is certain that $d_1 > d_2$, because $\Delta t_1 = \Delta t_2$

∴ ($tg\alpha_1 = \Delta t_1/d_1$) < ($tg\alpha_2 = \Delta t_2/d_2$). The unit is K/m. Therefore the number of temperatures drops per meter distance is called temperature gradient.

If we remember: "In poor insulation, temperature drops slowly, in good insulation, temp drops steeply", we can easily draw a temperature gradient curves as in Fig. 6-16(c) and (d). (c)'s insulation is on outside (winter cold side) and (d)'s insulation is on inside (warm side). They have the same thickness of concrete and insulation. Assume indoor temperature is 20℃, outdoor is −10℃. If take the same slow slope in concrete and the same steep slope in insulation, then we can draw temperature gradient curves, in the two sections, crossing the 0℃ horizontal lines at the "m" point respectively.

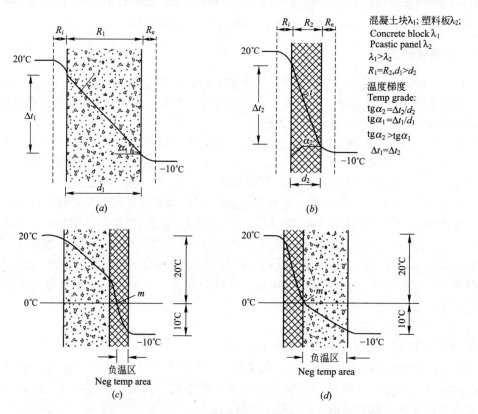

图 6-16 绝热层外包(冷侧)或内贴(暖侧)温度梯度比较

Fig. 6-16 Temperature gradients' comparison between insulation on outside (cold side) and insulation on inside (warm side)

6.3 分析 "Z" 与 "t_o"
6.3 Analyzing "Z" and "t_o"

Through analyzing we can learn an important concept: although material layers are the same, total resistances are equal, if the position of the material layers are different, the inside temperature distributions will be very different. In picture (c), wall with insulation on outside (cold side) has advantages as follows:

1) Negative temperature area is smaller so that there is less danger of condensation in the inner space;

2) In the section, mean temperature is higher. It's easier to keep indoor temperature stable.

It's suitable to continuously operating rooms such as dwellings, retirement villages…However the construction of insulation on outside is rather difficult. In picture (d), insulation on inside (warm side) is easy to construct. It's suitable to un-continuously operating rooms such as schools, shops, theaters and etc. If insulation is installed on warm side of the wall, there is less heat absorbed by wall and the indoor temperature rises quickly.

6.3 分析 "Z" 与 "t_o"

6.3.1 科学安排 "Z"

"Z" 是冬季散热时间，也是供热时间，以及夏季进热时间，同时也是供冷时间。为满足使用要求并节能，"Z" 的合理安排是：连续使用的房间如住宅等，连续供热（冷）；非连续使用的房间，非连续（间歇）供热（冷）如教室、商店、剧院等。进一步应向智能管理发展，做到不按季节只依气候（含突变性气候）控制 "Z"，达到更舒适，并更科学的节能。

6.3 Analyzing "Z" and "t_o"

6.3.1 Scientifically managing "Z"

"Z" is the time of winter heat loss or winter heating and summer heat gain or cooling. For the satisfactory use and energy saving, the reasonable management of "Z" should be: continuously heating (cooling) for continuously operating rooms such as dwelling houses and etc. and discontinuously heating (cooling) for discontinuously (intermittently) operating rooms such as classrooms, offices, shops, theaters, and etc. Intelligent control method should be further developed to manage "Z" according to the weather (incl. sudden change of weather) instead of to the seasons so as to achieve more comfort and more energy saving based on science.

6.3.2 人能改变室外气温 "t_o" 吗？

"t_o" 是室外气温，室外气温能人为改变吗？科学研究启发我们，有许多途径可改善局部地区的室外气温例如：

6 增效资源
6　INCREASING RESOURCES EFFICIENCY(IRE)

1）植树造林（含退耕还林），大力发展建筑绿化；

2）蓄水造湖，节水造池，保护湿地……

3）减少或停止排放有害物质，特别是 CO_2 大气温室气体；

4）延长日照。

图 6-17 为俄国拟用同步卫星反射太阳能延长日照时间，改善其高纬度区室外气温。黄昏来临，反射阳光仍可照亮设计范围的地面，次晨则提前照亮该地区。

对于单幢建筑，可采用外窗可反射板（图 6-18），延长室内日照。

6.3.2　Can the outdoor temperature "t_o" be changed manually?

"t_o" is outdoor air temperature. Can man change it? Scientific researches enlighten us there are many ways to improve an area's outdoor air temperature partially such as:

1) Afforestation (incl stopping farming, returning to foresting); greatly developing building landscape;

2) Storing water to build lakes, saving water to build pools, protecting wet lands…;

3) Reducing or stopping discharges of harmful substances, especially the atmospheric greenhouse gas CO_2;

4) To lengthen (prolong) sunshine

Fig. 6-17 shows that Russia is going to use a synchronous satellite to reflect solar energy to prolong sunshine time to improve the outdoor air temperature in high latitudes' area. When evening comes, the reflective sunlight can still reach the designed area. The reflective sunshine will rise earlier to shine the area next morning.

A building may have a reflective board installed outside the window to prolong indoor sunshine (Fig. 6-18).

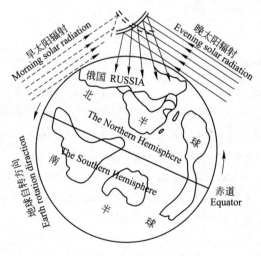

图 6-17　俄国研究同步卫星反射太阳能延长日照

Fig. 6-17　Russia had a study using a synchronous satellite to reflect solar energy to prolong the sunshine

6.4 分析"Q"与"$SRHG$"

6.4 Analyzing "Q" and "$SRHG$"

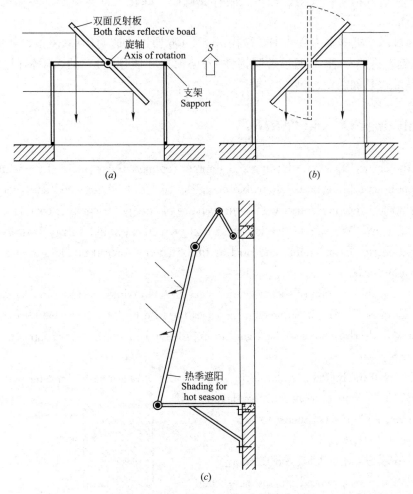

图 6-18 窗前可旋转反射板改善室内日照

Fig. 6-18　Improving indoor sunshine by a rotary reflective panel outside the window

(a)早晨室内提前日照；(b)黄昏延长日照；(c)夏季遮阳

(a)Indoor sunshine earlier than normal morning; (b)Prolonging indoor sunshine in evening;
(c)Summer sunshade

6.4 分析"Q"与"$SRHG$"

"Q"是冬供热夏供冷能源。当前，冬供热能源绝大多数仍是污染性、有限性的非可持续能：煤、气、油、柴、火电或核裂变电，夏空调第一能源也多是火电。应尽早大力发展太阳能、沼气、风能、水动力能、地热能、生物能以及核聚变能取而代之。

$SRHG$(solar radiation heat gain)为太阳辐射进热，冬季能减轻"Q"的负荷(见第5章)，夏季则增加"Q"负荷。此处着重讨论减少夏季太阳辐射进热，改善室内热环境。

关键点：

1) 设计好建筑平面形式、体形、建筑群的布置以及总图平面，获取好的 BSC、BVC、集中布置率(CLR)以及总平面图的周面比(P/A)(见 6.1.1～6.1.5)。

6 INCREASING RESOURCES EFFICIENCY(IRE)

2) 建筑构造 9 措施：反射、通风、阻存、遮阳、绿化、自然空调、多功能板、多功能窗及优化组合。

图 6-19 显示了夏季室内进热主要途径：Q_5、Q_6 通过非透明构件（屋顶、外墙）的太阳辐射进热与温差进热；Q_9 为通过透明构件（外墙窗户和屋顶天窗）的太阳辐射进热；Q_7、Q_8 为门窗缝隙进热和温差进热。

6.4 Analyzing "*Q*" and "*SRHG*"

"*Q*" is the energy for winter heating and summer cooling. Today, most of the winter heating energies are pollutant, finite, non-sustainable energies: coal, gas, oil, firewood and thermal power electricity or nuclear fission electricity. The first choice of energy for summer cooling is also thermal power electricity. We should develop solar energy, methane, wind energy, hydropower energy, geothermal energy, biotic energy and nuclear fusion energy instead of the unsustainable energies as early as we can.

SRHG is the abbreviation of solar radiation heat gain. In winter SRHG can decrease the load of "*Q*" (See Chapter 5). But in summer it increases the load of "*Q*". Here, we'll focus our discussion on how to decrease summer heat gain to improve indoor thermal environment.

Key points:

1) Design well the building's plan form, shape, layout of buildings and master plan to achieve good *BSC*, *BVC*, the rate of concentrative layout (*CLR*) and master plan's perimeter to area (*P/A*) ratio. See 6.1.1~6.1.5).

2) Nine measures of building construction: reflection, ventilation, resistance and storage, sunshade (sun shading), landscaping, natural air conditioning, multi-function panel, multi-function window and optimal combination.

Fig. 6-19 shows the main ways of indoor summer heat gain: Q_5 and Q_6 are the heat gain through the opaque members (roof, exterior walls) by solar radiation and by temperature difference; Q_9 is the heat gain through the transparent members (windows in exterior walls and skylights); Q_7 and Q_8 are the heat gain by gap penetration and temperature difference.

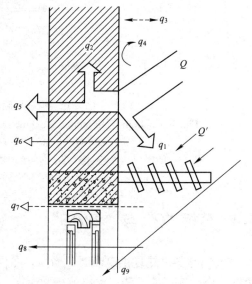

图 6-19 夏季太阳进热主要途径
Fig. 6-19 Main ways of summer heat gain

6.4.1 反射隔热

思考：在太阳辐射下用铝箔反射与用白色反射哪个隔热效果好呢？我们曾用镀锌铁皮作屋顶作过不同表面对太阳辐射热的隔热比较，见表 6-3。

6.4 分析"Q"与"SRHG"
6.4 Analyzing "Q" and "SRHG"

6.4.1 Heat insulation by reflection

Thinking deeply: which is better, to reflect solar radiation with aluminium foil or white surface? We have done a comparison of reflecting solar radiation with different surfaces on a galvanized iron sheet as a roof (Table 6-3).

不同表面反射辐射比较(实验)　　　　　　　　　　　表 6-3
A comparison of different roofing in reflecting solar radiation　　Table 6-3

表面温度(℃) Surface temp(℃)	白灰浆 Lime white	白漆 White paint	铝箔 Aluminium foil	红漆 Red paint	油毡 Asphalt felt	绿豆砂 Pea grit	绿豆砂刷白灰浆 Pea grit-lime white	白色中八厘石屑 White stone chip (size: 5-6mm)
外表面 Outer surface	39.3	41.8	43.5	50.4	56.5			
内表面 Inner surface	38.8	40.0	42.4	45.8	56.0	46.0	36.5	41.5

(Row "外表面" columns 6-8: 油毡不同保护层 / Different protective layer on asphalt felt)

此外我们还进行了下列实验：
1) 不同表面反射率比较，见表 6-4。
Furthermore, we have done the following experiments:
1) Different reflectance of different surfaces, see Table 6-4

不同表面反射太阳辐射比较(在现有房屋表面实测)　　　表 6-4
A comparison of the reflectance of different color walls　　　Table 6-4

不同表面 Different surfaces	刷白灰浆墙面 (4 年后) Lime white finish wall (after 4 years)	红砖墙 Red brick wall	青砖墙 Grey brick wall
表面温度(℃) Surface temp(℃)	40	51	53

2) 钢板门不同颜色反射率比较见表 6-5。
2) The comparison of the reflectance of different colors on steel sheet doors, see Table 6-5.

不同表面钢板门反射太阳辐射比较(现场实测)　　　　　表 6-5
A comparison of the reflectance of different colors on steel sheet doors　　Table 6-5

温度(℃) Temp(℃) \ 颜色 Color	深绿面漆钢板门(西晒) Dark green steel sheet door (Facing west)	刷白灰浆钢板门(西晒) Lime white steel sheet door (Facing west)	Δt(K)
外表面 Outer surface	62.5	44.0	18.5
内表面 Inner surface	61.0	43.0	18.0

表 6-6 为前辈们的实验资料。
Table 6-6, data of previous experiments

6 增效资源
6 INCREASING RESOURCES EFFICIENCY(IRE)

材料对热辐射的吸收率(反射率＝1－吸收率)　表6-6
Materials' thermal absorptance (Reflectance=1-absorptance)　Table 6-6

材料 Materials	长波 Long waves	短波 Short waves	材料 Materials	长波 Long waves	短波 Short waves
非金属黑色表面 Non-metal black face	0.90～0.98	0.85～0.98	发亮白涂料 Gloss white paint	0.85～0.95	0.15～0.20
红砖 Red brick	0.85～0.95	0.65～0.80	发亮铝漆 Gloss aluminium paint	0.40～0.60	0.20～0.30
黄或浅黄色砖或石表面 Yellow or light yellow brick or stone surface	0.85～0.95	0.50～0.70	镀锌钢材 Galvanized steel	0.20～0.30	0.40～0.65
米色砖；浅色表面 Cream color brick; Light color face	0.85～0.95	0.30～0.50	高抛光金属铬、铝 Highly polished chrome; aluminium	0.20～0.40	0.05～0.15

注：太阳光属短波辐射。

Note: Solar light is short wave radiation.

由上可知，白色(尤其白灰浆)反射太阳辐射比其他材料好。原因有二：一是白灰浆反射率高；二是其辐射散热强。归结起来，夏季用反射隔太阳辐射热，必须了解下述(1)、(2)、(3)辐射源与反射材料的特性：

From the above tests, we understand that white (especially the lime white) paint is better than other materials in reflecting solar radiation. The two reasons are: lime white has a high reflectance and a high radiant heat. To sum up, we must understand the characters of radiators and reflectors in order to reflect solar radiation for summer insulation. These are explained in the following points (1),(2),(3).

(1) 反射材料反射特性

图 6-20(a) 示出了几种典型表面的反射特性。在太阳光下白色表面比磨光铝反射还好，但在长波区，白色反射急剧降低，波长越长，反射率越低，波长大于 6um，白色与黑色反射率几乎无差别！高抛光铬、铝反射比白色和磨光铝还高，但建筑物不会用这些材料反射太阳热降低夏季进热，因为造价太高。

(2) 辐射源的发射特性

图 6-20(b)示出了几种辐射源发射特性。可以看出：辐射源表面温度越高，发射能量的高峰区越偏向更短的波段。

(3) 反射材料长波辐射散热的特性

例如白灰浆不仅反太阳辐射率高，而且长波辐射能力强(约为其他材料的五倍)，是一种效率高、造价低、易施工的好材料。但其严重缺点就是耐久性差。

必须指出，用反射减少夏季进热并不是可持续措施，只能在缺乏先进技术时权宜之计且用之。应尽快发展转换技术，将该太阳能转化为电力、绿化能、可用热能等，才是可持续之道。

(1) Reflectivity of reflective material

Fig. 6-20 (a) shows some typical surfaces' characters in reflection. In reflecting solar

6.4 分析 "Q" 与 "SRHG"
6.4 Analyzing "Q" and "SRHG"

radiation white surface is better than polished aluminium. But in long waves, the reflectivity of white surface drops rapidly. The longer wave the lower reflectivity. When the wavelength is longer than 6 uny the reflectivity between white and black are hardly different! The reflectivity of highly polished chromium and aluminium are higher than that of white paint and polished aluminium. But they are not being used in buildings to reflect solar heat for reducing summer heat gain, because of high cost.

(2) Emitters' radiative characters

Fig. 6-20 (b) shows some emitters' radiative characters. We may see that the higher temperature of emitter surface is the radiative energy summit is closer to the shorter wave area.

(3) Long wave radiative character of reflective materials

Example, lime white not only has a high reflectivity under sunshine but also has a strong long wave radioactivity (about five times of other materials). It's a good material of high efficiency, low cost and easy construction. Yet its serious shortcoming is poor durability.

We should point out that to reduce summer heat gain by reflection is a non-sustainable method. It's an interim measure due to lack of advanced techniques. We should develop conversion techniques to convert the solar energy into electricity, green energy, useful thermal energy, and etc as fast as we can. That's the sustainable way.

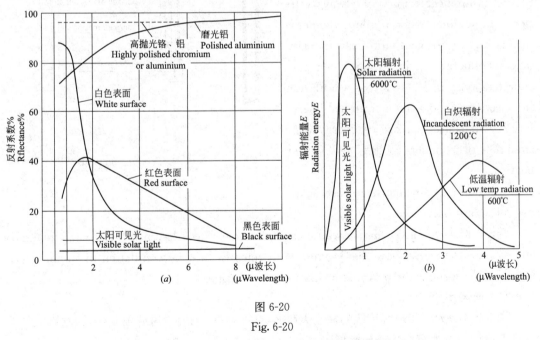

图 6-20

Fig. 6-20

(a)几种典型表面反射特性；(b)几种典型辐射源发射特性

(a) Reflectance of several typical faces；(b) Emissivity of several typical emitters

6.4.2 通风

通风散热两途径

1) 穿堂风。可利用蒸发制冷（或相变制冷）原理用穿堂风吹散室内的湿、热，如民居

6 INCREASING RESOURCES EFFICIENCY(IRE)

中的通风天井或公共建筑中共享空间的通风。穿堂风适用于湿热地区。

非湿热地区，穿堂风可用，但不总是有效的。夏季室外气温高时，穿堂风会使室内升温。冬季室外气温低，穿堂风会使室内气温低，以致夏降温、冬供暖能耗都将增加。

2) 构建通风道散热

(1) 通风墙

图 6-21(a)为湖南大学等设计的空心墙板通风墙。(b)为 Trombe 墙夏季白天通风散热，夜间将活动绝热层移开以利散热。

6.4.2 Ventilation

Two ways to release heat by ventilation are:

1) Cross ventilation、According to the principle of evaporation cooling (or phase change cooling), the cross ventilation can release indoor humidity and heat by ventilative yard in residential houses or communal spaces in public buildings. Cross ventilation is suitable to humid-hot region.

For non humid-hot region, cross ventilation is applicable, but not always efficient. In summer, when outdoor temperature is high, cross ventilation can increase indoor temperature. In winter, when outdoor temperature is low, cross ventilation can decrease indoor temperature. Therefore it will cause increase of energy consumption of summer cooling and winter heating.

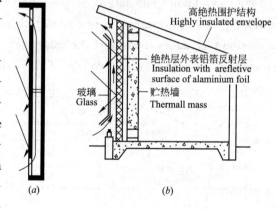

图 6-21 通风墙

Fig. 6-21 Ventilation walls

2) Construct air duct to release heat

(1) Ventilative wall

Fig. 6-21 (a) is a ventilative wall made by hollow panel, designed by Hunan University and others. (b) shows how the Trombe Wall releases heat by ventilation in summer daytime and by opening the movable insulation at nighttime.

(2) 通风屋顶

图 6-22(a)为广州地区传统的黏土大阶砖(400×400×30)通风屋顶；(b)为湖南大学等设计的钢筋混凝土空心板通风屋顶。

通风屋顶是双层构件：外层(热阻 R_f)和基层(热阻 R_b)，$R_b \gg R_f$ 好，即基层热阻越大越有利于绝热，使空气间层与室内温差所决定的夏进热少，冬失热也少。若空气层周边设有可开关通风孔，夏开、冬闭那就更有效了。以屋顶为例，基层有：承重层、绝热层、防水层、空气层，其上，即为外层，可用 3cm 厚黏土大阶砖或 3cm 厚水泥砂浆加筋板等构成。

6.4 分析"Q"与"SRHG"
6.4 Analyzing "Q" and "SRHG"

(2) Ventilation roof

Fig. 6-22(*a*) indicates a conventional ventilation roof built by large size clay bricks (400×400×30). (*b*) shows a reinforced concrete hollow panel ventilation roof, designed by Hunan University and others.

Ventilation roof is a double-layer member. It consists of external layer (thermal resistance R_f) and base layer (resistance R_b). $R_b \gg R_f$ is better, i. e. The higher base thermal resistance is the more insulation it has. The reduction of summer heat gain and winter heat loss depends on the temperature difference between the air space and room. If closable vents around the air space are installed, it would be more efficient to open them in summer and close them in winter. If taking roof as an example, base layer consists of load bearing structure, insulation, waterproofing and air space. Above air space is the external layer. It can be 3 cm thick large size clay brick or 3 cm thick reinforced cement board, and etc.

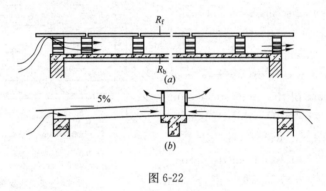

图 6-22

Fig. 6-22

(*a*)黏土大阶砖通风屋顶(广州地区); (*b*)钢筋混凝土空心板通风屋顶

(*a*)Large-size clay brick ventilation roof (Guangzhou region); (*b*)Reinforce concrete hollow slab ventilation roof

6.4.3 阻存

阻存是利用材料的热阻与贮热性能将太阳热能阻住并贮存起来,以减少室内进热(热期)或室内失热(冷期)。

图 6-23 为中国延安一组石窑洞。石窑、砖窑、土窑、靠山窑、下沉式窑,它们都有冬暖夏凉的性能。原因是周围都有深厚的土层,热阻大,热容量也大,夏季太阳辐射热都被阻住贮存起来。冬季向室内供热,这就是一种热延迟性或热惰性或热惯性。用"R"表示热阻,"S"表示贮热能力,D 表示热惯性,$RS=D$,优化用好"D"对节能减排很有利。

图 6-24 为一蓄水屋顶。利用水的大贮热容量取得夏降温良好效果。蓄水屋顶内表面27℃,比未蓄水屋顶内表面 33.3℃低 5.7℃!

6.4.3 Resistance and storage(RS)

RS means to use material's character to resist and store the solar heat to reduce indoor heat gain (in hot days) or heat loss (in cold days).

6 增效资源
6 INCREASING RESOURCES EFFICIENCY(IRE)

Fig. 6-23 is a group of stone caves in Yanan city. Stone cave, brick cave, loess cave, hillside cave and sunken cave all have the character of keeping warm in winter and cool in summer. Because there is very thick soil surrounding these caves. They have large heat resistance and storage capacities. Solar radiation in summer can all be resisted and stored. In winter, the stored heat will heat the room. This is the heat delay or thermal inertia. Generally "R" represents thermal resistance, "S" represents heat store capacity, D represents thermal inertia. $D=RS$. Optimally managing "D" will benefit energy saving and reducing pollutants' discharge.

图 6-23 延安石窑群

Fig. 6-23 Stone caves in Yanan, China

Fig. 6-24 shows a water storage roof relying on large heat storage capacity of water has achieved a good result of summer cooling. The temperature of the inner surface under the roof pool was 27℃ while the inner surface of roof without water was 33.3℃. The former was 5.7℃ lower than the latter.

图 6-25 是 50%锯末与 50%黄土混合 30cm 厚的种植屋面，是我们自然绿化实验的一例。我们不浇水、不种植，在西安年降雨量 640mm 的自然状态下，飞来的种子，竟然树草丛生，有的树已高达 4m 多，主根系横布大部分屋面面积。夏季室外气温 35℃，内表面只有 26℃。

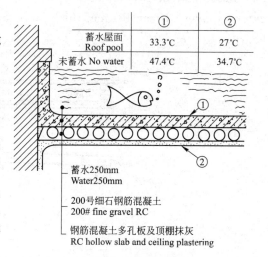

图 6-24 屋顶水池夏季隔热

Fig. 6-24 Roof pool for summer insulation

（重建大实验）

(Experimented by Chongqing University of Arch)

Fig. 6-25 shows a landscaped roof with a 30cm planting layer of 50% sawdust mixed with 50% loess. This is an example of our natural landscaping experiment. We never water and plant it. But under the natural condition of annual rainfall of 640mm in Xi'an region, the flying seeds landed on the roof. Trees and grass have overgrown. Some trees are more than 4 meters high. Their main roots have grown across and occupied most area of the roof. In summer, when outdoor air temperature is 35℃, the inner surface of the ceiling is only 26℃.

6.4.4 遮阳

遮阳主要是阻止夏季太阳直射热通过窗户入室。

图 6-26 示出了三种水平遮阳构造：

（a）水平遮阳实心板，宜用于朝赤道方向（在北半球朝南，南半球朝北）。这种水平实

6.4 分析 "Q" 与 "SRHG"

6.4 Analyzing "Q" and "SRHG"

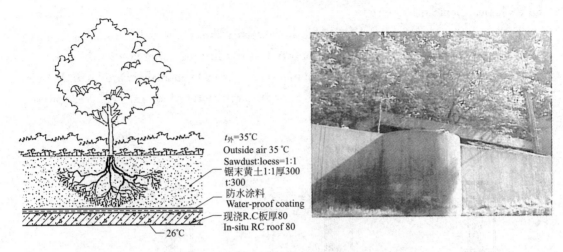

图 6-25 锯末黄土种植屋面

Fig. 6-25 Sawdust-loess planted roof

心遮阳板的缺点是，会阻止贴墙面热气流散热。一般墙面在太阳直晒下，表面温度会高于室外气温（我们曾测到灰砖墙面温度高达 53℃）当时室外气温 35℃。

(b) 水平遮阳百叶克服了上述缺点，既能遮夏季太阳直射，又有利于贴墙气流散热。但其构造较复杂，造价会较高。

(c) 水平可种植遮阳板，只要铺 8～10cm 轻质种植层（蛭石或珍珠岩，或锯末黄土……）就可种牵牛花、月季花等多种花卉。夏遮阳、冬日照，并有生态效益。

东、西向窗户则宜用垂直遮阳。因为东、西向太阳高度都低，水平遮阳效果甚微。图 6-27(a)为垂直遮阳板与水平通风百叶相结合；(b)为带窗的锯齿形墙。垂直遮阳构件也可绿化。

6.4.4 Sunshade

Sunshade mainly stops direct solar heat in summer into rooms through windows.

Fig. 6-26 shows three horizontal sunshade structures:

(a) Solid horizontal sunshade board. It's suitable to the orientation facing equator (in the Northern Hemisphere, facing south; in the Southern Hemisphere, facing north). The shortcoming of the solid horizontal sunshade board is that it can prevent the warm air adjacent to the wall from releasing heat. Under solar radiation, the surfaces' temperatures of common walls are higher than that of the air (We have measured a grey brick wall surface temperature reaching 53℃ while the air temperature was 35℃!)

(b) Horizontal sunshade shutter can overcome the above shortcoming. It can shade direct solar radiation in summer and allow the adjacent air to blow off heat. But its construction is rather complex and expensive.

(c) Horizontal plantable sunshade board can bear a 8 to 10cm light planting layer (vermiculite or pearlite, or sawdust mixed with loess…). Then we may plant morning glory, Chinese rose and many other flowers. We can get summer sunshade, winter sunshine and ecological benefits.

East and west facing windows should have vertical sunshade. Horizontal sunshade is hardly useful on east and west facing windows because in these orientations the solar altitudes are low. Fig. 6-27 (a) shows a combination of vertical sunshade boards with horizontal ventilation shutters. (b) shows the saw tooth wall with windows. The vertical sunshade elements can be landscaped too.

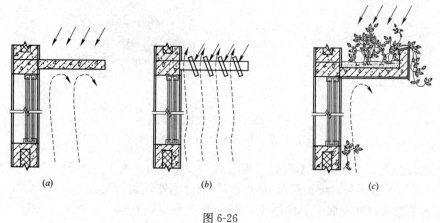

图 6-26

Fig. 6-26

(a)水平钢筋混凝土实心板；(b)水平遮阳百叶；(c)水平可种植遮阳板

(a) Horizontal sunshade made of solid reinforced concrete board; (b) Horizontal sunshade shutter; (c) Horizontal plantable sunshade board

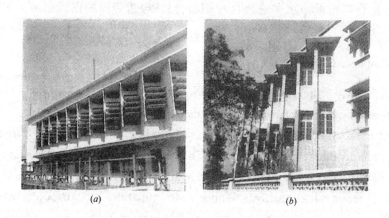

图 6-27 夏遮阳

Fig. 6-27 Summer sunshade

6.4.5 绿化

我们 20 多年实验证明：对建筑物散水、窗台、阳台、墙面、屋面进行绿化，技术不难，花钱不多，不仅建筑热环境得到改善，而且对美化环境、净化空气也都大为有利（详见 8 建筑绿化）。全球建筑绿化将大有利于平衡大气温室效应及消除气候过度变暖。

6.4 分析 "Q" 与 "SRHG"

6.4 Analyzing "Q" and "SRHG"

6.4.5 Landscaping

More than 20 years' experiments of ours have proved that to plant on building's apron, windowsill, balcony, wall and roof is not difficult in technology. It's low cost and can not only improve the thermal environment but also greatly beautify the environment and purify the air (Detail in 8, Greening of Buildings). The global building landscaping will greatly benefit balancing AGE and eliminating COW.

6.4.6 自然空调

此处，自然空调是指利用水、土或其他材料的热惯性对入室空气进行温湿度调节。例如窑洞土层温度夏季比室外气温低，冬季比室外气温高，在其中设通风道，使室外空气经该风道入室，就可得到夏降温，冬增温以及调节湿度的效果。无污染，无加剧热岛负效应缺点，无破坏臭氧层的危害(详见 4.3.5 及 7.3.5)。

6.4.6 Natural air conditioning (NAC)

Here, NAC means to use water, soil or other material's thermal inertia character to adjust the temperatures and humidity of the air coming into rooms. For example, in soil layer surrounding a cave, its temperature is lower than that of outdoor air in summer and higher than that of outdoor air in winter. If a vent duct is installed through the layer to let outdoor air go into the room via the duct, it will be cooled in summer and heated in winter and its humidity adjusted. No pollution, not aggravating thermal island negative effect, no damage to ozonosphere (Detail in 4.3.5 and 7.3.5).

6.4.7 多功能构件

将屋面、墙面、阳台，以及窗台、遮阳板、雨篷、散水变为将太阳能转换成热、电、绿化能的多功能构件，必将使建筑的生态可持续性大大增强。

6.4.7 Multi-function members

To make roof, wall, balcony and windowsill, sunshade board, canopy, apron multi-function members to convert solar energy into heat, electricity and green energy would greatly increase building eco-sustainability.

6.4.8 多功能窗

传统概念：窗功能只有三个字：风、光、景。新概念：窗功能有 6 条：风、光、景、热、电、空调。"风"就是利用窗户通风换气。"光"就是利用窗户自然采光，并获取日照。"景"就是通过窗户观景，并使居者与外界自然信息得到交流。"热"、"电"、"空调"含义是：①窗户是直接得热太阳房必要构件；②科研已发现，透明光电转换材料，可使窗户自然采光、日照并发电；③一间窗户小温室(可种植)就是一间自然空调的小空间。

但是，窗户一直是冬失热、夏进热最多的构件，同时也是隔声薄弱构件。

6 INCREASING RESOURCES EFFICIENCY (IRE)

6.4.8　Multi-function window

In conventional concept, window's functions are summarized into three words: ventilation, lighting and view. In new concept, window's functions have six parts: ventilation, lighting, view, heating, electricity and air conditioning. "Ventilation" means through windows to get air exchange. "Lighting" means through windows to receive day lighting and sunshine. "View" means through windows to look out and exchange information between the occupants and the nature. "Heating", "electricity" and "air conditioning" mean that: ①Windows are the essential members of direct gain solar buildings; ②Scientific discovery on transparent light-electricity converting materials allows windows to provide day lighting, sunshine and electricity; ③A small window greenhouse (plantable) is a small natural air conditioning space.

But, window is still a member from which the room loss most heat in winter and gain most heat in summer and has poorest sound insulation.

改善措施：

1) 加强气密性，用密封条封缝

图 6-28(a)图为双扇单玻窗夹有一空气层的复合窗；(b)单扇双玻夹一空气层。(a)复合窗可用于已有房屋（原有单玻窗加一扇单玻窗就成为双扇单玻窗了）。(b)单扇双玻宜用于新建房，构造简单、气密性好。图 6-29 为一国外双玻窗例，密封条为氯丁橡胶，各部件嵌接采用"↑"箭头形弹簧，拆装方便。

密闭空气层 3~5cm 的热阻比 12cm 黏土砖墙或 23cm 混凝土墙热阻还大。但密闭空气层热阻并不随厚度增加而增大，并与热流方向和气温有关。热流向下，对流传热停止，热阻增大；夏季气温高，空气分子运动加快，热阻减小；空气间层一侧贴铝箔，热阻大增。详见表 6-7。

Improvements:

1) Increasing air tightness with weather stripes to seal joints.

Fig. 6-28(a) shows a composite window with an air space between two single glazing windows; (b) shows a double-glazing window with an air space. (a) Composite windows are suitable to existing buildings (it only needs to attach a new single-glazing window to the old one.) (b) Double-glazing windows suit to new buildings. It is simple construction and has good air tightness. Fig. 6-29 shows an example of a double-glazing window in a foreign country. The weather stripe is made of neoprene. The joints between parts are jogged by the "↑" arrow-headed springs. They are easy to be assembled or dissembled.

The thermal resistance of a 3 to 5 cm closed air gap is larger than that of a 12 cm thick clay brick wall or a 23 cm thick concrete wall. But the thermal resistance of a closed air gap is not increased with the increasing of its thickness. Its resistance is related to heat flow direction and air temperatures. When heat flows downward, the convection will be stopped and the resistance will be increased. In summer, when air temperature is high, air molecule move quicker. The resistance will be decreased. With an aluminium foil attached to one side of the air-gap, the resistance will be greatly increased. Detail in Table 6-7.

6.4 分析"Q"与"SRHG"
6.4 Analyzing "Q" and "SRHG"

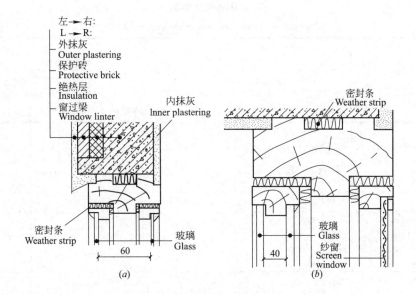

图 6-28 窗户气密性构造

Fig. 6-28 Air tightness of windows

(a)双扇单玻璃;(b)单扇双玻窗

(a)Two single glazing windows; (b)A Double-glazing window

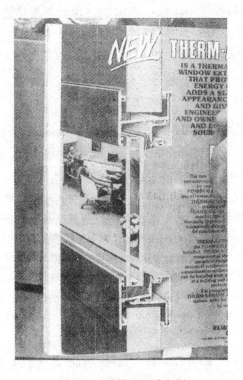

图 6-29 国外—双玻窗例

Fig. 6-29 A double-glazing window in foreign country

6 增效资源
6 INCREASING RESOURCES EFFICIENCY(IRE)

密闭空气层热阻 表 6-7
Thermal resistance of sealed air gap(m² K/W) Table 6-7

空气层状况 Air gap situation		冬 Winter						夏 Summer					
		空气层厚度(cm) Air gap thickness (cm)						空气层厚度(cm) Air gap thickness (cm)					
		0.5	1	2	3	4	5	≥6	0.5	1	2	3	4	5	≥6
热流方向 Heat flow direction	↓	0.10	0.14	0.17	0.18	0.19	0.20	0.20	0.09	0.12	0.15	0.15	0.16	0.16	0.15
	↑	0.10	0.14	0.15	0.16	0.17	0.17	0.17	0.09	0.11	0.13	0.13	0.13	0.13	0.13
	→	0.10	0.14	0.16	0.17	0.18	0.18	0.18	0.09	0.12	0.14	0.14	0.15	0.15	0.15
单面铝箔空气层 Aluminium foil against one side of the air gap	↓	0.16	0.28	0.43	0.51	0.57	0.6	0.64	0.15	0.25	0.37	0.44	0.48	0.52	0.54
	↑	0.16	0.26	0.35	0.40	0.42	0.42	0.43	0.14	0.20	0.28	0.29	0.30	0.30	0.28
	→	0.16	0.26	0.39	0.44	0.47	0.49	0.50	0.15	0.22	0.31	0.34	0.36	0.37	0.37

2) 减少接缝

图 6-30 示出了一住宅窗(窗口 1.5m 高×1.2m 宽)。设室内气温 $t_i=20℃$，室外气温 $t_o=-10℃$，隙缝宽平均1mm，渗透气流速率1m/s(实际调查：隙缝宽均大于1mm，渗透气流速率多大于1m/s)。该窗 12 块 3mm 厚普通玻璃，全部接缝总长为 25.2m。经计算：全天 24 小时渗透失热为 85338.5kJ；通过玻璃温差失热为 14210kJ；非透明部件失热 4574kJ。该窗户全天 24 小时总失热为 104122.5kJ，空气渗透失热比：

$\frac{85338.5}{104122.5}\times100\%=82\%$。由此可见，隙缝密封和减少接缝多么重要！

2) Reducing joints

Fig. 6-30 shows a window in a dwelling (window opening is 1.5m high by 1.2m wide). If indoor air temperature $t_i=20℃$, outdoor air temperature $t_o=-10℃$, the average window gaps width is 1mm, penetrating air speed=1m/s (practical data: gap width>1mm; penetrating air speed>1m/s). The window has 12 pieces normal glass of 3mm thick. The total length of all joints' gap is 25.2m. Calculations indicate that during 24 hours, the heat loss by penetration is 85338.5kJ; the heat loss by temperature difference through glass is 14210kJ; the heat loss through non-transparent (opaque) parts is 4574kJ. During 24 hours, the window's total heat loss is 104122.5kJ. The rate of air penetrating heat loss is (85338.5/104122.5)×100%=82%. From here we can see how important it is to seal the gaps and reduce joints!

3) 热镜

普通玻璃有"透短吸长"的特性。科研发现：有些透明的半导体含有氧化锡或氧化铟或氧化锌，具有"透短反长"的特性，即透过短波辐射，如太阳辐射，反射长波辐射(长波热镜效应)。这些热镜材料为增强热工效能提供了新条件。

图 6-31 示出了氧化铟锡热镜特征。

6.4 分析"Q"与"SRHG"
6.4 Analyzing "Q" and "SRHG"

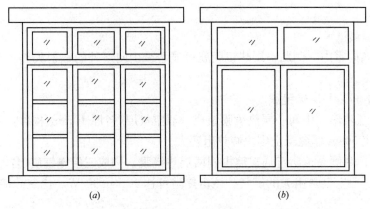

图 6-30 减少窗户接缝

Fig. 6-30 Reducing window joints

(*a*)传统住宅窗；(*b*)改进后

(*a*)Conventional dwelling window；(*b*)Improved

3) Heat mirror

The common glass has a character of "Transmitting short wave radiation while absorbing long wave radiation". Science has discovered that some transparent semi-conductors containing oxide of tin or indium or zinc have the character of "Transmitting short wave radiation such as solar radiation while reflecting long wave radiation" (long wave heat mirror effect). These heat mirror materials provide us a new possibility for increasing thermal efficiency.

Fig. 6-31 shows the nature of the heat mirror of indium-tin oxide.

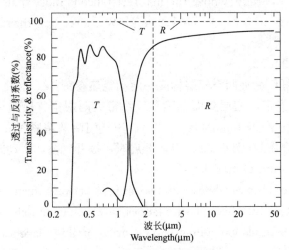

图 6-31 氧化铟锡热镜特性

Fig. 6-31 The nature of the heat mirror of indium-tinoxid

对 0.3~1.0μm 短波辐射有高透过率；对>2.5μm 长波辐射有高反射率；虚线所示为理想热镜特性：

对>2.5μm 长波辐射全反射(100%反射率)，对≤2.5μm 短波辐射全透过(100%透过率)

The nature of the heat mirror of indium-tin oxide. High transmissivity within 0.3~1.0μm short wave radiation.

High reflectivity to the long wave (>2.5μm) radiation. Ideal heat mirror: reflects entire long wave

(>2.5μm) radiation, transmits entire short wave (≤2.5μm) radiation.

The dotted line shows the idea heat mirror nature

6 增效资源
6 INCREASING RESOURCES EFFICIENCY(IRE)

热镜用法：

(1) 内表面热镜膜

在普通玻璃向室内一侧贴一层热镜膜就可显著减少向室外的辐射失热，并保持较好的太阳辐射透过率。

(2) 双玻中间装一层热镜膜

① 两层普通玻璃，中间一层热镜膜组成一双空气层(各厚10mm)构件，冬季热阻可达 $0.4m^2K/W$，比50cm厚混凝土砌块墙热阻还大。

② 国外例：两侧高透明普通玻璃中间两层热镜膜，构成三层空气层(各厚10mm)，空气层内充以比一般空气热阻大的氩气。该窗热阻可达 $1.0\ m^2K/W$，比三砖墙(74cm厚)热阻还大。

How to use heat mirror：

(1) Inner surface heat mirror film

To adhere a heat mirror film (membrane) to the glass inner surface (facing room) can obviously reduce outwards radiant heat loss while keep better solar radiation transmissivity.

(2) One heat mirror film between double-glazing

① To install one heat mirror film between normal double-glazing is equal to construct a double air gap (each thickness of 10mm) member. Its winter thermal resistance can reach $0.4m^2K/W$, more than that of a 50 cm thick concrete block wall.

② A foreign example：Two heat mirror films installed between two high transmissivity normal glazing produced a member with three air gaps (each thickness of 10mm). The gaps are then filled with argon, whose thermal resistance is more than that of the air. This window's thermal resistance can reach $1.0\ m^2K/W$, more than that of a three clay brick wall (74cm thick).

4) 活动绝热层(图 6-32)

三层双面铝箔塑纤织品夹两空气层构成一活动绝热卷帘。

图 6-32 所示绝热卷帘由三层双面铝箔塑纤织帘夹两层空气层组成，四周密闭，可由大气压自由充气入空气层(厚5mm)(图 6-32a)，用气门活塞堵紧气孔。该帘四周与窗/门周边用塑料毛刺布啮合密封(图 6-32c、d)。开启时，拨开气塞，手动或电动卷起即可。

4) Movable insulation (Fig. 6-32)

Three layers of fiber-plastics fabrics with two air gaps between them become a movable insulation rolling curtain. The three fiber plastics fabrics have aluminum foils on both sides.

Fig. 6-32 shows an insulation rolling curtain made of three layers of fiber-plastics' fabrics, with aluminium foils on both sides and two air gaps between them. Its surrounding is sealed and the air gaps can be freely filled with air by the atmospheric pressure (each air gap 5mm thick) Fig. 6-32 (a), with spiracle-piston plug up the air hole. The curtain's surrounding against the surrounding of window/door is sealed airtight with thorny plastics strips (Fig. 6-32(c)(d)). To open it, pull out the piston then roll up the curtain artificially or electronically.

5) 改善隔声

上述加强气密性，减少接缝，增加空气层，以及设活动绝热层都会同时改善隔声。此

6.4 分析"Q"与"SRHG"
6.4 Analyzing "Q" and "SRHG"

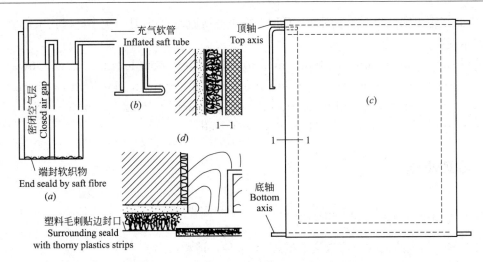

图 6-32 三层双面铝箔夹两空气层活动绝热帘
Fig. 6-32 Movable curtain made of three layers of fiber-plastics fabrics with two air gaps. The three fiber plastics fabrics have aluminium foils on both sides.

外,采用双层玻璃时,室内一侧用 3mm 普通玻璃,外侧用 5mm 安全玻璃,两玻厚度不一,不易共振传声。或外侧玻璃用弧形玻璃,冬季可多得些阳光,同时也会增强隔声。窗户温室(图 6-33)宜用推拉窗,外侧玻璃斜安装对隔声与冬季多得阳光均有利。当面向赤道时,该玻璃倾角∠β=当地地理纬度+15°~20°为宜;当为东或西向时,∠β=当地地理纬度+25°~30°为宜。背阳面窗户也可做成温室种植喜阴植物。

5) Improving sound insulation

The above-mentioned strategies of strengthening airtightness, reducing joints, increasing air cavity and installing movable insulation, all can improve sound insulation. Besides, when using double-glazing with 3mm inner normal glass and 5mm external safety glass, the different thickness is not easy to resonantly transmit noise. Alternatively, curved glazing can be used outside to receive more sunlight in winter and increase sound insulation. Window greenhouse (Fig. 6-33) is suitable to use sliding window. Its external sloping glazing is good for sound insulation and to receive more sunlight in winter. When the window greenhouse is facing equator, the glazing's gradient angle ∠β=Local geographic latitude+15°~20° is better; when it's facing east or west, the ∠β=Local geographic latitude+25°~30° is better. Window not facing the sun can be also made into a greenhouse to plant shade-liking plants.

上述多种双玻窗(含窗户温室)玻璃,白天隔声量

图 6-33 窗户温室
Fig. 6-33 Window greenhouse

可达35dB或更多些。晚上盖上活动绝热层，隔声可增加到50dB甚至更多些(与24cm黏土砖墙隔声相近)。

The sound insulation of above-mentioned various double-glazing windows(incl window greenhouse) can reach 35dB or more during daytime. At night, with movable insulation covered, the sound insulation can be increased to 50dB or more (it's near to the sound transmission loss of a 24cm clay brick wall).

6.5 本章小结

启思式 $Q_{SHG}^{WHL} = \dfrac{t_i - t_o}{R_o} FZ \mp SRHG$

真奇妙，启发我们找出增效资源措施50余条，请看：

1) "F"：The less$(BSC)(BVC)(SLB)(P/A)$ the more benefiting IRE's10 benefis

2) $R_o \begin{cases} R_p = R_i + R_w + R_e \cdots\cdots 围护结构实有热阻 \\ R_{o \cdot min \cdot N} = \dfrac{t_i - t_o}{[\Delta t]} R_i \cdots\cdots 1980s\,设计围护结构低限热阻 \\ R_{o \cdot min \cdot ES} = \dfrac{100}{100-n} R_{o \cdot min \cdot N} \cdots\cdots 节能热阻 \end{cases}$

懂得上述三公式就会按节能要求设计围护结构，如：
(1) 重型复合墙(增阻微增重)：密闭空气层，铝箔(一空、二暖、三净)，无冷桥联接，无砂浆联接，预制复合墙板……。
(2) 轻型复合墙——发展方向
(3) 多层材料复合墙——热阻相等、材料布置不同，内部热工效果大异(图6-16；6-34)

6.5 Conclusion of the chapter

Enlightening formula $Q_{SHG}^{WHL} = \dfrac{t_i - t_o}{R_o} FZ \mp SRHG$

Enlightening formula, has wonderfully inspired us to find more than 50 items about IRE's measures. Let's see：

1) "F"：The less$(BSC)(BVC)(SLB)(P/A)$ the more benefiting IRE's 10 benefits

2) $R_o \begin{cases} R_p = R_i + R_w + R_e \cdots\cdots \text{Practical envelope resistance} \\ R_{omin \cdot N} = \dfrac{t_i - t_o}{[\Delta t]} R_i \cdots\cdots \text{Minimum resistance for designing envelope during 1980s} \\ R_{o \cdot min \cdot ES} = \dfrac{100}{100-n} R_{o \cdot min \cdot N} \cdots\cdots \text{Energy saving resistance} \end{cases}$

Understanding the above 3 formulas we may design envelope according to energy saving demands, such as：
(1) Heavy composite walls (increasing resistance with micro-weight increased): closed air gap, aluminium film (one air gap, winter warm side, clean), no thermal bridge

6.5 本 章 小 结
6.5 Conclusion of the chapter

joints; no mortar joints, precast composite panels….

(2) Light composite walls—developing direction

(3) Multi-layered composite walls having equal thermal resistance, different positions of materials, will achieve very different inner thermal effects. (Fig. 6-16; 6-34)

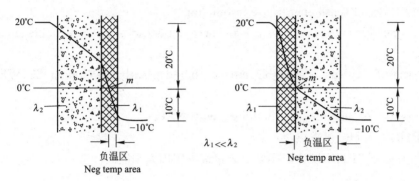

图 6-34

Fig. 6-34

(4) 影响"λ" 6 因素：密度、湿度、内部微孔状况、热流方向、分子/原子/电子传热的活跃性、时间。了解它们对选材、生产、施工均有指导意义。

(5) 轻型延性(柔性)结构与轻型复合构造相结合的建筑是重大灾害下保护生命财产最佳建筑类型之一。

3) "t_o"：如何用科技方法改变局部室外气温？

(1) 植树造林；

(2) 大力发展建筑绿化；

(3) 蓄水造湖；

(4) 节水/导流造池；

(5) 保护湿地；

(6) 延长日照(高科技，普科技)。

4) Q_{SHG}^{WHL} 广义建筑用能

(1) 可持续能源(清洁能源)取代非可持续能源(污染能源)；

(2) 城乡(特别大城市)应大办太阳能、风能、沼气能联网；

(3) 大力发展自然空调太阳房。

(4) 6 factors affecting "λ": density, humidity, inner micro porous state, heat flow direction, heat conducting activity of molecule, atoms & electrons, time. To understand them will guide for material selection, producing and construction.

(5) Light ductile (soft) structure combining with light composite construction's building is one of the best building types protecting lives & properties from the serious calamities.

3) "t_o": How to change "t_o" partially by sci-tech?

(1) Foresting;

(2) Greatly developing building landscape;

6 增效资源
6 INCREASING RESOURCES EFFICIENCY(IRE)

(3) Storing water to make lakes;
(4) Saving and dredging water to build pools;
(5) Protecting wet land;
(6) Lengthening sunshine (high tech, normal sci-tech).

4) Q_{SHG}^{WHL}: Broad sense energy-use in building

(1) Sustainable energy sources (clean energy) instead of un-sustainable energy (pollutant energy);
(2) Cities & villages (esp. big cities) should greatly developing SE, WE, ME network;
(3) Greatly developing solar buildings with NAC.

5) SRHG：太阳辐射进热：
(a) 做好应用 SRHG 设计，详见 5.2 建筑中应用太阳能；
(b) 夏防 SRHG
(1) 优化 $(BSC)(BVC)(BL)(P/A)$；
(2) 反射抓三性；
(3) 通风两途径；
(4) 阻存多厚重；
(5) 遮阳要巧新；
(6) 绿化妙搭配：季青常青搭配，高乔低灌搭配，美化除害搭配；
(7) 自然空调潜能大（调温、调湿、净化空气、杀菌、消除热岛负效应、安全疏散……）；
(8) 多功能构件办法多：
① 太阳能多形式转换：
太阳能→热、光、电、绿化能、热惯性能；
② 节能效率多途径提高：
● 加强气密性（用密封条堵缝）；
● 减少接缝，热镜，铝箔夹空气层，窗户温室（节能、隔声均提高）。
(9) 优化组合选最优。
例：两方案：a)，b)：
设：$(BSC)_a$ 好于 $(BSC)_b$，$(BVC)_b$ 好于 $(BVC)_a$，$(BL)_a$ 好于 $(BL)_b$，$(P/A)_b$ 好于 $(P/A)_a$，那么 $(BSC)_a (BVC)_b (BL)_a (P/A)_b$ scheme is the best.

★ 21 世纪，人类若能充分利用太阳能、水能、风能、氢能、沼气能就可能做到清洁能源自足，并防止争夺能源的纷争与战争，为和谐世界创造了最重要条件之一。

5) SRHG: Solar radiation heat gain;
(a) Design well to use SRHG, detail in 5.2 Solar Energy-use in Buildings;
(b) Preventing SRHG in summer
(1) Optimizing (BSC) (BVC) (BL) (P/A);
(2) Reflective insulating SRHG should consider 3 characters;
(3) Ventilation has two ways;

6.5 本章小结
6.5 Conclusion of the chapter

(4) Resistance & storage insulating SRHG usually rely on thick-heavy materials;

(5) Shading sunshine should be ingenious;

(6) Landscaping needs wonderful combinations: Deciduous greens combining with evergreens, tall trees combining with short bushes, beautification combining with eliminating harms;

(7) NAC has more potentials (adjusting temperature & humidity, cleaning air, killing germs, eliminating thermal island negative effect, safe escape…);

(8) Multi-function members have many ways:

① Multiform of solar energy conversion:

Solar energy→energy of heat, light, electricity, landscaping and thermal inertia;

② Energy saving efficiency increased by many methods:

To strengthen airtightness by weather strip;

Reducing joints; Heat mirror; window greenhouse (ES and SI all increased).

(9) Select the best by optimizing combination

Example: There are two schemes: a), b):

If: $(BSC)_a$ *is better than* $(BSC)_b$, $(BVC)_b$ *is better than* $(BVC)_a$, $(BL)_a$ *is better than* $(BL)_b$, $(P/A)_b$ *is better than* $(P/A)_a$, *then* $(BSC)_a (BVC)_b (BL)_a (P/A)_b$ scheme is the best.

★ During 21st century if human beings can fully use the solar energy, water energy, wind energy, hydrogen energy and methane energy, the world may be self-sufficient on clean energy and avoid the disputes and wars fighting for energy. It would create one of the most important conditions for a harmonious world.

7 掩 土 建 筑
7 EARTH SHELTERED BUILDINGS

7.1 引论

这里所指掩土建筑乃是部分或全部被土覆盖的建筑。

从旧石器时代人们就已经住着天然洞穴和人工地下建筑。其原因常常是为了抵御室外坏气候，另外也是为了省出地面，以便从事农业或举行礼仪活动，还可能是为了御敌。无论是为了什么，人们已体会到地下空间能提供比室外舒适得多的微气候。例如中国北部的黄土已被证明适于地下居所，从而可空出地面作农业用地。

7.1 Introduction

Earth sheltered building in this book means the building is partly or wholly sheltered (covered) by earth.

Since the Old Stone Age, people have been living in natural caves and underground buildings. The reason is often to resist outdoor bad climate or to leave land for agriculture or amenity actions, or for defence. Whatever it is for, people had experienced that underground spaces can supply a better microclimate than outdoor spaces do. For example, the loess in north China has been proved very suitable to construct underground dwellings in order to leave the ground for agriculture.

在干旱区，黄土特别适于农业。在热、旱气候下，地表难于耕作时，黄土却有一定保持内部湿度的能力。用作建造窑洞居所时，黄土有足够的强度构成居室空间（利用拱形结构），而且易于挖掘。这些地下村落（彩图 7-13a）已存在了几千年，实是这种系统实用性和有效性的证明。

In arid zones, loess is very suitable for agriculture. Under hot and arid climate when the ground is hard to cultivate the loess still has capacity to retain moisture. When loess is used to build cave dwellings, it has enough strength to construct living spaces (using arched structure) and is easily excavated. These underground villages (Color picture7-13(a)) have existed for several thousand years, which do become the proof of practicality and effectiveness of this system.

地下建筑的主要优点来自土壤的热工性质。厚重的土覆盖层所起的热惯性作用使其覆盖的空间内温波很小。不同的土壤导热性也不同，主要取决于物理成分、化学组成、土的含湿度以及土壤密度。土层温度受到太阳辐射、大气温度以及土壤热惯性的作用会产生由表及里逐层减幅的波动，到一定深度将消失波动，保持一定的恒温称为恒温线。例如，中

7.1 引　　论
7.1 Introduction

国西安地区(黄土地区)年恒温线在地下 15m 左右处,恒温为 16℃。随地理纬度、土壤性质不同,年恒温线深度也不同,一般在 4～16m 范围。年恒温线的深度可人工方法改变。例如地面种草植树、加绝热层或对土加湿。

The main advantages of underground buildings result from the soil thermal characteristics. The effect of the thermal inertia of thick soil cover makes low temperature amplitude in the internal space. Different soil has different conductivity dependent on its physics, chemical components, humidity and density. The effects of solar radiation, air temperature and the soil thermal inertia cause the temperatures in the soil cover to fluctuate with an amplitude gradually lower and lower from its surface to a certain depth until the amplitude disappears to a constant temperature called Constant Temperature Line (CTL). For example in Chinese Xi'an region (loess region) the annual CTL is at 15m underground and its constant temperature is 16℃. The annual CTL will vary with differences of geographic latitude and soil character. In general the annual CTL depth is within 4～16m. The annual CTL's depth can be artificially changed by planting grass and trees, adding heat insulation and increasing soil humidity, and etc.

场地选择

将掩土房设置在低平地段的缺点是:易遭洪水淹没;沙尘暴掩埋;眼前视景狭窄,增加了闭塞感;另外,通风和空气循环也可能受到限制。

掩土房位于山顶则有某些优点:通风好;视景好;自然采光好;排水方便。但是,地下居所建于山顶,施工困难,上下交通不方便,花钱多。

将掩土房屋建在斜坡场地上,优点较多。气候更适宜,自然采光好,视野开阔,便于组织排水,通风条件好,闭塞的不良感觉可减到最少。按等高线设计,出入通道也方便。但这种场地比平地场地修筑道路投资要多。

Site selection

When earth sheltered houses are located in low flat area the shortcomings are: inundated by flood; buried by dust storm; narrow field of vision and increased feeling of depression. Furthermore, ventilation and air circulation might be limited.

Earth sheltered houses on hilltops can get some advantages such as: good ventilation; good field of vision; good natural lighting; easy drainage. But earth sheltered houses on hill tops will be difficult to construct, inconvenient access ups and downs, and more expensive.

Earth sheltered houses on sloping sites have more advantages: better climate; good day lighting; wide view; easy drainage; good ventilation conditions; low depressive feeling. If it is designed according to the contours, the entrances and exits can also be convenient. However the cost of road development on sloping site is more than that of a flat site.

与地面非掩土建筑比,掩土建筑有如下优点:
1) 节能节地;
2) 微气候较稳定;
3) 防震、防风、防尘暴、防火、防辐射;
4) 绝热、隔声、隔空气污染良好;
5) 洁净(医学菌落试验已证实);

7 掩 土 建 筑
7 EARTH SHELTERED BUILDINGS

6) 安静；

7) 新陈代谢平缓（人体生理试验已证实），这是一个长寿因素；

8) 较安全，如防盗，应急避难；

9) 外部维修面少；

10) 有利于生态平衡及保护原自然地景。

Compared with above ground non-earth sheltered buildings, the earth sheltered ones have the following advantages:

1) Energy saving and land saving;

2) Stable microclimate;

3) Quakeproofing, wind proofing, dust-storm proofing, fire proofing and radiation-proofing;

4) Good insulation of heat, sound and air pollutions;

5) Cleanness (The medical colony test has proved);

6) Quiet;

7) Gentle metabolism (Human physiologic test has proved), this is a factor for longevity;

8) Better security e. g. against burglary, emergency shelter;

9) Less maintenance of building exterior surface;

10) Greatly benefiting eco-balance and keeping original landscape.

除浅层地下空间（入地＜15m）及地面掩土建筑外，中深层地下空间最主要的难点可用五个字概括，即水、火、风、光、逃。

水：即施工时的地下水处理问题及使用期的排水问题；

火、逃：逃离难，灭火难；

风：地下空间自然通风条件较差，必须有强大的机械通风或精心设计的被动系统保证通风换气；

光：地下空间自然采光条件差。

Except shallow underground spaces (below ground＜15m) and above ground earth sheltered buildings, for middle and deep underground spaces we may use five words to sum up their issues. These are: "water", "fire", "ventilation", "lighting", "escape".

"Water": It is difficult to deal with underground water during construction and drainage in usage;

"Fire", "Escape": Hard to escape and put out the fire;

"Ventilation": In underground space natural ventilation condition is poor. It needs careful designed strong mechanical ventilation or a passive system to ensure the ventilation and air exchange;

"Lighting": In underground space natural lighting condition is poor.

科学家们正致力研究几何光学的引光系统及光导纤维的引光系统。图 5-26，图 5-28 是两个实验例。他们还在研究建立地下自己的可持续系统，如：人造太阳，废物和废水更新再用以及地下自然空调等。

Scientists are researching geometric optic light-catcher system and fiber optic light guide system. Fig. 5-26 and Fig. 5-28 are two experimental examples. They are also researching to build underground self-sustainable system such as: man-made sun, renewal and reuse of waste material and water and underground natural air conditioning, etc.

7.2 国外掩土建筑实例

当代许多国家特别是发达国家已经建造了一大批掩土建筑与地下空间,例如:

英、法:该两国人口加起来不及我国的 1/10,但其现代地下空间的开发利用却很发达,除大城市的地铁和地下建筑外,通过多佛尔海峡(加来海峡),连接英国多佛尔和法国敦刻尔克两国的海底通道也早已正式营运。

7.2 Examples of Earth Sheltered Buildings Internationally

Today in many countries especially in developed ones large number of earth sheltered buildings and underground spaces have been built, for example:

In UK and France: the population together is less than one tenth of the population of China. But their modern underground spaces have been greatly developed. Besides the subways and underground buildings in cities, the underwater tunnel through the Strait of Dover (Pas de Calais) connecting UK Dover and France Dunkerque is already in use.

巴黎的多层地铁及地下超级商场都是世界闻名的地下空间。

美国:有地下住宅、实验室、图书馆、数据处理中心、高级计算机中心、100多所地下学校等。尤其一些有名的大学,为了保护地面景观,都建有地下图书馆。

In Paris multi-leveled subways and underground supermarkets are world famous underground spaces.

In USA, many dwellings, laboratories, libraries, data centers, super computer centers, and more than 100 schools have been built underground. Many famous universities have their libraries built underground to protect aboveground landscape.

日本:已成为发展大型地下空间的先导国家之一。日本已有 30 多座城市建有地下购物城(Underground Shopping City 称为"Chikagai"——地下街)。有的地下街有 300 多家商店,每天可容纳 80 多万顾客。

Japan has been one of the leading countries for developing underground spaces. More than 30 cities have built "Underground Shopping Cities" called "Chikagai". Each can contain more than 300 shops allowing 0.8 million shoppers everyday.

加拿大:如蒙特利尔地下综合区,占地面积 80 多万平方米。有 2 个火车站、9000 个停车场、3 座百货公司、4 座豪华宾馆、8 个剧院、和多处餐馆、办公房、和一所大学。这个多用途地下综合区可同时容纳 50 万人。

In Montreal, Canada, there is an underground complex occupying more than 0.8 million m^2 underground area containing 2 railway stations, 9000 car parks, 3 department stores, 4 luxury hotels, 8 theatres, and many restaurants, office buildings and an university. This multi-purpose underground complex can be occupied by 0.5 million people at one time.

7 掩 土 建 筑
7 EARTH SHELTERED BUILDINGS

俄罗斯：莫斯科除原有环形与放射网络地铁外，在前苏联时期，已有20余所院校参与了多功能地下空间的开发规划，将地下建筑与地铁连系起来，空出地面作为户外体育和其他使用。莫斯科地铁每天运输量平时约为800万人次；节假日达1000万人次以上。

In Moscow, Russia, besides the existing circular and radial subway networks, during the time of former CCCP more than 20 universities had participated the planning of the multi-use underground spaces. They planned to develop underground buildings connecting with subways so as to leave ground area for outdoor sports and other uses. The average daily passengers volume of Moscow subway is about 8 million person-trips and in holidays over 10 million per day.

瑞士：50%以上电站建在地下。

In Switzerland, more than 50% of power stations are built underground.

匈牙利：布达佩斯市只有200多万人口就有3层(红、黄、蓝)三色地铁，四通八达。

In Budapest, Hungary, the population is only a little more than two million. Yet there are three-level subways in three colors: red, yellow and blue.

在军事建筑中，许多国家特别是发达国早已发展了大批现代地下空间。

In military architecture, many countries especially developed ones have already developed a lot of modern underground spaces.

以下各图是一些国家地下和地上民用掩土建筑实例：

彩图7-1为美国明尼苏达大学地下空间中心的地上部分视景，最深处达地下30余米。地上部分是水墙集热太阳能房(图5-27)。这是一座世界闻名的掩土太阳能建筑，并有塔楼利用几何光学系统为地下房间进行引光、引景实验。

The following pictures are the examples of underground and aboveground earth sheltered civil buildings in several countries:

Color picture 7-1 shows the view of ground part of the Underground Space Centre, Minnesota University, USA. The deepest part of the center is more than 30m below ground. The above ground part is a water wall solar building (Fig. 5-27). This is a word famous earth sheltered solar building with a tower using geometric optical system to capture light and views for the subterranean rooms.

彩图7-2为法国巴黎一地下超级市场入口，地下商场空出的地面则可作为公园等用。彩图7-3为地面公园一例。

Color picture 7-2 shows an entrance of an underground supermarket in Paris, France. Supermarkets located underground spaces can leave ground for parks, etc. Color picture 7-3 is an example of an aboveground park.

彩图7-4(a)为法国HAGETMAU市市政府入口处视景。这也是一座世界闻名的掩土太阳房。彩图7-4(b)为屋顶绿化与下沉式温室部分视景。人们称该市政府为花园市政府。市政府所有办公房间(市长室、各职能科室用房、电脑管理室、礼堂、活动圆桌会议厅、小会议室等)均设在地下，除屋顶花园外，还在入口处设有一个下沉式花园(下沉式温室)。从图中还可看到草坪屋顶上的圆采光罩。

Color picture 7-4(a) shows the sights around the entrance of HAGETMAU City gov-

7.2 国外掩土建筑实例
7.2 Examples of Earth Sheltered Buildings Internationally

ernment. This is also a world famous earth sheltered solar building. Color picture 7-4(b) shows a partial view of the roof planting and a sunken greenhouse. People call the government a "Garden government". All the offices (mayor's room, various offices, computer rooms, assembly hall, movable round-table meeting room, small meeting room, etc.) are built underground. Besides the roof garden, a sunken greenhouse has been built near the entrance. From the picture we can also see the round skylights on the grassed roof.

彩图7-5(a)为美国明尼苏达州首府明尼阿波利斯市数据控制中心,是一座大型掩土太阳能建筑。图7-5(b)为其剖面示意。图7-5(c)为其毗连日光间剖面示意,建筑面积23039m², 共5层(含顶楼),钢筋混凝土结构,东、西、北三面覆土种草,绿化至第三层。第四层无覆土,直接自然采光。南面为一大型毗连日光间,面积1560m²(彩图7-5d)。其内建有一个容量为230m³水池作为贮热(冷)库。采暖热量取自日光间太阳能以及回收的计算机运行散热、照明器具散热和人体散热。夏季采用机械制冷。

Color picture 7-5(a) shows the Control Data Center building in Minneapolis, the capital of Minnesota, USA, a large earth sheltered solar building. Fig. 7-5(b) is its schematic section. Fig. 7-5(c) indicates schematic section of the attached sunspace. The building has a floor area of 23039m², five stories (including the penthouse) and reinforced concrete structure. The east, west and north faces are earth sheltered and planted from the bottom to the third floor. The fourth floor has no earth cover and is exposed to receive natural lighting. The south face is a big attached sunspace covering an area of 1,560m² (Color picture7-5d). Within the sunspace there is a thermal storage pool with a total capacity of 230m³. The winter heating is supplied by the sunspace solar energy and recycled radiant heat from computer operation, lighting and output from human activity. In summer one chiller operates for cooling.

该建筑的关键是采用了掩土和毗连日光间。掩土不仅可种草绿化,还有良好的热惯性作用,减少了室内的冬失热,夏得热。毗连日光间不仅提供了供暖的热量和建筑南侧房间的补充天然采光,同时也增加了建筑的美感。

The key point of this building is the utilization of earth cover and an attached sunspace. Earth cover not only can be grassed but also has good thermal inertia in reducing winter heat loss and summer heat gain. The attached sunspace not only supplies heating

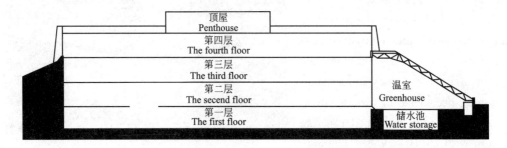

图7-5(b)明尼阿波利斯市数据控制中心掩土太阳能建筑剖面示意

Fig. 7-5(b) Shows the earth sheltered solar building's schematic section of the Control Data Center in Minneapolis

7 掩土建筑
7 EARTH SHELTERED BUILDINGS

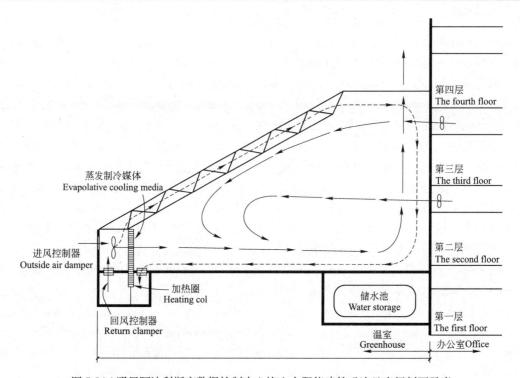

图 7-5(c) 明尼阿波利斯市数据控制中心掩土太阳能建筑毗连日光间剖面示意

Fig. 7-5(c) Shows schematic section of the attached sunspace of an earth sheltered solar building, the Control Data Center in Minneapolis

but also the ancillary day lighting for the south side rooms while increases the aesthetic quality of the building.

图 7-5(e)为明尼阿波利市一信息中心地下空间接待室。该中心对外公开服务，可提供最新信息与各方面历史资料。该中心为单层钢筋混凝土结构。室内人工照明，温、湿度一直较稳定，电脑及其他机电设备工作稳定，文字资料及信息软件保存质量好。缺点是通风，照明耗能大，人在其中与外界自然信息交换太少。

Fig. 7-5(e) shows the reception room of an underground information center in Minneapolis. This center is open to public and supplies latest information and historic documents in various areas. It is a single-story reinforced concrete structure with indoor artificial lighting. Temperature and humidity are stable. Computers and other electronic and mechanical equipments work stably. Documents and software are well stored. The

微笑接待 Smiling welcome

图 7-5(e) 明尼阿波利斯市信息中心地下空间接待室

Fig. 7-5(e) Shows the reception area of the Underground Information Center in Minneapolis

7.2 国外掩土建筑实例
7.2 Examples of Earth Sheltered Buildings Internationally

shortcomings are its high energy consumption for ventilation and lighting and little information exchange between people stay inside and outside environment.

彩图 7-6(八代克彦博士提供)为 1990 年 5 月,在中、日双方探讨 21 世纪城市更新与地下空间开发研讨会上,日本一建筑师提出的 21 世纪日本地下城设想方案之一。它是一座几千米直径的巨大的钢筋混凝土筒体,深入地下几十米,分层建筑多种类型建筑群。在筒体底部相邻的蜿蜒行进的水平筒体内是滋浮列车,其他筒体交叉网络为水、电、暖、通(风)等管道空间。

Color picture 7-6 (Source: Dr Yashiro) shows one of the design schemes for the 21st century underground city in Japan suggested by a Japanese architect at the Forum on 21st Century City Rehabilitation and Development of Underground Spaces, which was held between architectural researchers from China and Japan, in May 1990. The design is a very big RC tube structure of several km diameters buried under the earth several tens of meters. Different type of buildings would be built on different levels. At the bottom, pipes housing magnetic-suspension trains are extending next to the bottom of the tube. The other pipe networks supply water, electricity, heat, ventilation, and etc.

本方案作为探讨人类生存空间的扩展,是一个有创意的大胆设想。

This is an imaginative and courageous vision of developing human living space.

彩图 7-7 为联合国教科文组织(UNESCO)设在巴黎总部的地下二层办公楼群,共六个下沉式开敞庭院,作为该办公楼群采光井。访问时,官员作答:"确有冬暖夏凉的热工特性带来的节能优点,室内微气候也较稳定"。屋顶草坪广场对竖立在广场后面的该总部办公高楼提供了一个开阔的绿色视景。

Color picture 7-7 is a view of the underground 2-story office buildings of UNESCO's Headquarters in Paris. There are six sunken open courtyards forming the light wells for providing day lighting for the office buildings. During our visit, the officer's answer is: "there is a real advantage of energy saving for keeping warm in winter and cool in summer. The indoor microclimate is also more stable". The open square with lawn on the roof provides a wide green outlook for the UNESCO Headquarter's high rise office building standing behind.

彩图 7-8 为法国巴黎一沿街掩土住宅视景。用户以覆土种花草。朝向院落的一面是向阳面,为各家入口。各家装有太阳能热水器与太阳能电池板。居民说,感受冬夏均很舒适(辅助空调几乎未开动过),防噪音、防路尘比地面楼房好。

Color picture 7-8 is a view of the earth sheltered dwellings along a street in Paris, France. The residents have soil covered area with grass. The entrance of each dwelling faces the court and the sun. Each home has its own solar hot water collector and solar cell panel. The users spoke to the author that they have experienced very comfortable conditions in winter and summer (few has operated ancillary air conditioners). Sound insulation and dustproofing are better than in an aboveground non-earth sheltered building.

彩图 7-9 为澳大利亚中部某卫星地面中继站掩土建筑。地面只一座接收微波天线,其余用房、仪器设备、电脑器件等全部设在地下,在钢筋混凝土顶盖上,保持原有的地形地景。由于地下空间温、湿度较稳定,电脑器件、机器设备运行良好。这是一个很成功的掩土建筑,

7 掩土建筑
7 EARTH SHELTERED BUILDINGS

地面景观保持原样，同时却争得了一个很大的地下工作空间。（资料来源：格林兰德博士）

Color picture 7-9 shows the earth sheltered building of a satellite relay station in the center region of Australia. On the ground, there is only a receiving aerial. The other rooms, instruments, computers, and etc are all constructed underground. The RC roof is earth covered to keep the original geography and landscape. Due to the stable temperature and humidity in underground space, computers and instruments operate very well. This is a successful earth sheltered building. The original ground landscape has been kept while provided a large underground working space. (Source: Dr Greenland, Australia)

彩图 7-10 为日本几所地下街实例：彩图 7-10(a) 为日本神奈川县川崎地下街一景；彩图 7-10(b) 为该地下街天窗采光视景；彩图 7-10(c) 为另一个地下街入口。日本 30 多座城市的地下街网，有的日流动量可达 80 万人次以上，对缓解地面的拥挤现象起了很大的作用。（资料来源：日本加藤羲夫先生）

Color picture 7-10 shows some built examples of Japan's "Chigakai" Underground Shopping City. Color picture 7-10(a) is a view of a "Chigakai" in Kawasaki, Kanagawa and Color picture7-10(b) is the skylight of the same Chigakai, Color picture7-10(c) is another "Chigakai" entrance. In Japan more than 30 cities have "Chigakai". Some have the daily capacity to accommodate more than 0.8million person-time. It has greatly decreased crowdness on the ground (Source: Mr. 加藤羲夫).

彩图 7-11 为法国巴黎罗浮宫地下博物馆出入口，建筑师为世界著名建筑师贝聿铭，结构设计为法国一著名工程师（通过他的女儿 KARINE 送给夏云一套彩色幻灯片，可惜我们忘了他的名字）。透明金字塔形结构对罗浮宫地面古典建筑遮挡最少，对于地下博物馆又可提供一定的太阳光能和热能。

Colour picture 7-11 is the entrance of the underground museum of the Plais du Louvre, Paris. It is designed by the world famous architect I. M. Pei and a famous French structure engineer (We are sorry to forget his name. His daughter Miss Karine has sent Xia Yun a set of color slides). This transparent pyramidic entrance minimizes the obstruction to the view of the Louvre's classical architecture aboveground while supplies certain amount of day lighting and heating to the underground museum.

彩图 7-12(a) 为法国一掩土别墅外景。图 7-12(b) 为该别墅内景。室内光线柔和亲切，一年四季室内微气候均舒适宜人。

Color picture 7-12 (a) shows external view of an earth sheltered villa in France. Fig. 7-12b The indoor microclimate is very comfortable at all seasons.

图 7-12(b)
Fig. 7-12(b)

7.3 中国窑洞

7.3.1 中国窑洞的特点

中国窑洞是材料最少、建造最简、历史最长、住人最多的掩土建筑。20世纪80年代以来，日本、澳大利亚、美国、英国、法国、瑞典、瑞士等国有关专家、教授、大学生、研究生纷纷来华考察，研究中国的窑洞。日本东京工业大学八代克彦博士三年多，跑遍了陕、甘、豫、晋、青及新疆等地区进行调研，他用风筝携带照相机拍的下沉式窑洞村落照片(彩图7-13a)已成为世界闻名的图像资料。日本邮政曾作为明信片发行。法国波尔多建筑学院Jean—poul LOUBES 让·波·鲁波教授等与作者们也进行过很好的合作，研究中国窑洞。英国剑桥大学 Helen Mullingan 海伦·茉莉根博士、原西安医科大学张宜仁教授与夏云曾经紧密合作，一起赴农村考察，共同书写多篇论文，至今仍留下良好的回忆。我国有关大专院校、研究单位也对窑洞作了许多很有成效的研究与实验。

7.3 Chinese Caves(Yao dong)

7.3.1 The characteristics of Chinese caves

Chinese caves are earth sheltered buildings that use the least materials, built by the simplest construction, have the longest history and still have largest number of users. Since 1980s, the relative experts, professors, students and postgraduates from Japan, Australia, USA, UK, France, Sweden, and Switzerland have come one after another to China to observe and study Chinese caves. Dr. Yashiro, from Japan Tokyo Industry University, has stayed in China for more than three years, traveling through many provinces such as Shaanxi, Gansu, Henan, Shanxi, Ningxia and Xinjiang Region for researches. With a camera carried by a kite, he took a picture of a sunken cave village (Color picture 7-13a). This picture has become a world famous photographic material. Japan Post has published the picture as a postcard. Prof Jean—Poul LOUBES and others from France Bordeaux Architecture School and the authors have cooperated well in studying Chinese caves. Dr Helen Mullingan, Cambridge University, UK, Chinese Prof Zhang Yiren (Xi'an Medical University) and Xia Yun had close cooperation together in studying Chinese caves, investigated villages and wrote papers…and have left a good recollection about the collaboration. Many other interested Chinese colleges, universities and institutes have done many useful research and experiments on Chinese caves too.

人类必将继续以现代科技手段开发地下空间。要重返地下就必须寻源并创新。科研工作者从研究中国、埃及、突尼斯等国掩土建筑的历史中，在有关建筑设计原理与方法方面得到了以下启发：

Human beings will certainly continue to develop underground spaces using modern science and technology. But to develop underground space further we must study the history

7 掩 土 建 筑
7 EARTH SHELTERED BUILDINGS

and be creative. By studying the history of earth sheltered buildings in China, Egypt and Tunisia, and etc, scientists have been inspired in following areas in building design principles and methods:

1)"冬暖夏凉"的掩土建筑是对极端气候:严寒、酷热、尘暴等作斗争的良好防护体。

1) Earth sheltered buildings' ability to keep warm in winter and cool in summer is good defence against extreme climates: severe cold, extreme hot, dust storms, and etc.

2) 古代掩土建筑是就地取材、技术简易、造价最经济的范例。

2) Ancient earth sheltered buildings are good examples of using local materials, simple technology and minimum cost.

3) 在场地选择、朝向定位、利用坡地、低洼地,精用平地等,先辈们因地制宜创造了许多有用的设计与建造方法。

3) In terms of selecting site, choosing orientation, using sloping site and low land and carefully using flat lands, and etc, our ancestors have created many suitable design and construction methods according to local conditions.

4) 中国下沉式黄土窑洞为开发浅层地下空间提供了有用的依据,如:

① 下沉式窑洞地面一般均在地表下 6m。黄土地层深 6m 的温度年波动正好与室外空气年波动出现 180°的相位差。即室外最冷月份(1~2月)温度最低峰-8~-12℃;该 6m 深地层温度却处于最高峰处(16.5℃左右);室外最热月份(7~8月)温度最高峰 38~40℃甚至更高,该 6m 深地层温度却处于最低峰处(14.5℃左右)。这就是窑洞冬暖夏凉的主要原因之一。

② 像下沉式窑洞这样的浅层地下空间不仅具有良好的地温及较厚土层热惯性的围护作用,而且日照、通风、采光、居住者对外界信息的感受与在传统地面建筑无大差别。

③ 中、深层地下空间会遇到地下水、日照、通风、采光、紧急疏散及与外界信息相通等难题。浅层地下空间则无此麻烦。

④ 设计、建造浅层地下建筑比建造地面高层建筑及中、深层地下建筑较简易。

4) Chinese sunken (pit-type) loess caves have provided useful evidence for developing shallow underground spaces. For example:

① the floor of pit-type cave is generally 6m below the ground. At this depth of loess, the annual temperature wave is just 180° out of phase with the outdoor air annual temperature wave i. e. in the coldest months (January to February), when outdoor air is at the lowest temperatures of $-8\sim-12$℃, the 6m deep loess is at the highest temperature of about 16.5℃. In the hottest months (July to August), when outdoor air is at the highest temperatures of $38\sim40$℃ or even more, the 6m deep loess is at the lowest temperature about 14.5℃. That is one of the reasons that the caves can be warm in winter and cool in summer.

② The shallow underground spaces such as pit-type caves not only have better soil temperatures and thermal inertia of thick soil as an envelop but also have similar conditions of sunshine, ventilation, day lighting and the users' experiences of communication with

7.3 中国窑洞
7.3 Chinese Caves (Yao dong)

external environment to conventional ground buildings.

③ In construction of middle depth or deep underground buildings, there are problems of groundwater, sunshine, ventilation, day lighting, urgent escape and information exchange between users and outside. But in shallow underground spaces there is no such trouble.

④ To design and build shallow underground building is simple and easy compared to the design and construction of high-rise buildings or middle depth to deep underground buildings.

5) 古代掩土建筑，尤其中国窑洞是蕴含当地文化及建筑文脉的范例。

5) Ancient earth sheltered buildings, especially Chinese caves, are good examples preserving local cultural and architectural contexts.

7.3.2 中国窑洞类型

(1) 根据场地形式分

1) 下沉式窑洞(彩图 7-13a，b)。

2) 靠山窑洞(图 7-14a、b 及彩图 7-14c)。

7.3.2 Types of Chinese Caves

(1) By site forms

1) Pit-type (Sunken) caves (Color picture 7-13a, b)

2) Hillside caves (Fig. 7-14a, b and Color picture 7-14c)

每一下沉式院落一户。每一院落四面。东、西、南三面每面各开两孔窑，北向开一孔窑和一出入坡道。七孔窑洞，人居三孔，家畜二孔，厨房一孔，一孔贮藏杂用，紧邻家畜窑洞设旱厕。院内常设有水井一口，贮藏院落雨水。有的地方也有大院，一院多户。

One family usually occupies one sunken courtyard. Each courtyard has four facades. In the east, west and south facades users can excavate two caves in each side and in north facade excavate one cave and one ramp as the entrance. Among the seven caves three are used for people, two for livestock and one for cooking. Adjoining the livestock caves there is a dry toilet. The courtyard usually has a well for collecting rainwater. In some places, a big sunken courtyard may house several families.

图 7-14(a)、(b)为陕西礼泉县烽火中学靠山窑洞师生宿舍。该窑洞群是当地居民挖掘黄土山体而成，为安全在内表面及窑洞脸面贴衬黏土砖护面。国内外人士多次参观，有两优点给来访者印象特别深刻：一是建靠山窑洞省出平地作农业，同时该窑洞又可得到好的风、光、景；二是将挖出的土填平山沟，形成坡道出入口，同时又堵截了山洪冲毁田、房。

Fig. 7-14(a)、(b) show the dormitory rooms for teachers and students in Fenghuo Middle School in Liquan County, Shaanxi Province, China. The cave groups were built by excavating the loess mountain by local people. For safety, the caves' inside and outside surfaces were finished by clay bricks. Many Chinese and international visitors have visited it. Two features particularly have impressed them: one is the use of hillsides to build caves to save flat land for agriculture while the caves can have good breeze, light, and view.

7 EARTH SHELTERED BUILDINGS

The other is to fill the gully with excavated soil to form a slope as the school entrance while stop the mountain torrents from damaging farmland and houses.

图 7-14
Fig. 7-14
(a)靠山窑洞群师生宿舍；(b)女生在宿舍门前的合影
(a) Hillside caves as dormitory rooms for teachers and students；(b)Schoolgirls photographed in front of their bedroom

彩图 7-14(c)为陕西延安大学靠山石窑洞群（师生员工宿舍），全为当地岩石开凿的石块砌拱填土而成，是世界上已知的最大的石窑洞群。

Color picture 7-14 (c) is the hillside stone cave group (as teachers and students' dormitory rooms) of Yanan University, Shaanxi, China. All caves were arched by stone blocks excavated from local rock mountain, and finally back filled with soil. This stone cave group is the biggest one known in the world.

(2) 根据结构材料分

1) 黄土窑洞（彩图 7-13a、b）——直接在黄土山侧或下沉院落四周开挖而成；
2) 石砌窑洞（彩图 7-14c）——从岩石山开凿石块，砌拱回填土，夯实即成；
3) 黏土砖拱窑洞——先砌砖拱，再回填土而成（图 7-15）；
4) 土坯拱窑洞——先砌土坯拱，再回填土；
5) 岩石窑洞——天然岩石洞穴或人工直接开凿石山而成；
6) 砂—卵石窑洞（图 7-16）——从砂石山体直接开挖而成。

(2) According to structural materials

1) Loess caves (Color picture 7-13a, b) are formed by directly excavating into loess hillsides or ground to form sunken courts；

2) Stone arch caves (Color picture7-14c) are formed with stone blocks excavated from rock mountain, then arched and backfilled with rammed earth；

3) Clay brick arch caves are formed by firstly laying brick into vault then backfilled with soil (Fig. 7-15)；

4) Adobe arch caves, an adobe vault is built first then backfilled with soil；

5) Rock caves are natural rock caves or formed by excavating into rock；

6) Sand-pebble caves (Fig. 7-16) are formed by directly excavating into sand-pebble hills.

7.3 中国窑洞
7.3 Chinese Caves (Yao dong)

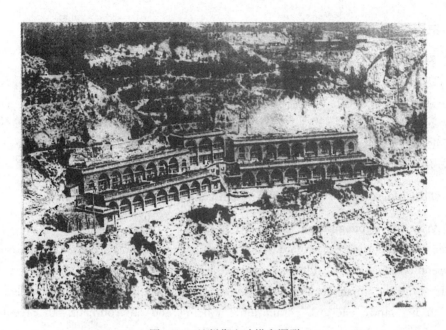

图 7-15 兰州靠山砖拱窑洞群
Fig. 7-15 Hillside brick arch caves in Lanzhou

中国窑洞全都是拱形结构，但各地窑洞门窗装饰无不源自地方文脉。彩图 7-17 即一例：红纸花饰窗及双鹤珠门帘陪衬着古朴的窑脸，充分显示出地方色彩。

Chinese caves are all arch structures. But the decoration of their doors and windows originates from their local context. Color picture 7-17 is an example: the window is decorated with red paper flower and the door is covered with a curtain made of plastic beads to form a pattern of two cranes. The simple and unsophisticated cave face perfectly illustrates the local color.

图 7-16 延安砂—卵石窑洞
Fig. 7-16 Sand-pebble caves in Yan'an

窑洞有深厚的土层包围，不仅有适合人、畜居住的冬暖夏凉的条件，还是一个良好的天然冷藏库。彩图 7-18 中小女孩正手拿土豆告诉人们，该小窑洞就是她家的天然节能冷藏库。库内有土豆、红薯、鸡蛋、萝卜等等。

Caves with very thick soil as an envelope, not only provide the conditions to keep warm in winter and cool in summer but also a good natural cool room. In Color picture 7-18, the little girl handling the potato tous us that this small cave is just their natural energy saving cool room in which potatoes, sweet potatoes, eggs and radishes and etc. are stored.

7 掩 土 建 筑
7 EARTH SHELTERED BUILDINGS

7.3.3 中国老窑洞缺点与综合治理

(1) 为什么年轻人不爱住窑洞？

从先进国家现代地下空间的开发看出人类以现代科技手段重返地下的强大势头，从缓和土地、能源、人口、建房、环境之间错综矛盾看出重返地下的重要性。但是，在我国近些年来，窑区人民弃窑盖房的风气却极盛行。原因有二：一是旧窑洞确实存在不少缺点（如塌窑死、伤人），二是年轻人认为住窑洞是住"寒窑"，被人看作贫穷、低下。调查中也发现，几乎所有老年人被窑洞的冬暖夏凉的舒适性吸引，爱住窑洞，不愿搬家。

7.3.3 Old Chinese caves' shortcomings and comprehensive management

(1) Why don't young people like to live in caves?

From the modern underground space developed in advanced countries we can feel a strong trend of re-entering underground space by modern science and technology. We can also understand the importance of using underground space to resolve the complex contradictions among land, energy, population, housing and environment. In recent years, however Chinese people living in caves have increasingly abandoned their caves to build aboveground houses. There are two reasons: one is that the old caves in fact have many shortcomings. Another reason is that the youngsters consider living in cave as living in "poor caves". This makes others consider them poor and inferior. Through investigation we discovered that almost all old people like living in caves and do not wish to move to aboveground dwellings. They are attracted by the comfort of caves keeping warm in winter and cool in summer

各类窑洞以黄土窑洞分布最广，而缺点又最严重，概括起来有五个字：塌、潮、暗、占（占地多）、塞（闭塞、通风不好）：

1) 塌顶：由于雨季雨水渗入顶土及溯击窑脸土，使顶土及窑脸坍塌。1985年，陕西全省不完全统计，雨后塌窑8万多孔，发生人畜伤亡事故。

Among various cave types, the distribution of loess caves is the widest. Yet their problems are most serious. The issues may be summarized in five points: collapse, damp, dark, taking up more land and poor ventilation.

1) Collapsing roof: During the rainy season, rainwater penetrates cave roof and wash off soil from cave façade. It causes roof and cave façade to collapse eventually. In 1985, according to incomplete statistics, Shaanxi Province had more than 80 thousand loess caves collapsed after rainy period and caused injuries and deaths to people and livestock.

2) 潮湿：潮湿天主要发生在夏季。当室外空气带着高温、高湿入室后，空气温度迅速降低，比地面建筑室内温度低8～10℃，故窑内空气相对湿度迅速增高（可达90%以上）。窑洞内表面温度比其室内气温更低，故而产生表面凝结水，以及墙脚及角落部位霉变。

2) Damp: Most wet days are in summer. When outdoor air with high temperature and high moisture comes into caves, the air temperature will be rapidly dropped to 8°～10°C

7.3 中国窑洞
7.3 Chinese Caves (Yao dong)

lower than that in the aboveground houses. Then, in the cave, the air's relative humidity will be quickly increased and may reach higher than 90%. Because the surface temperature of cave rooms are lower than that of the indoor air. This causes surface condensation and mildewing at the bottom and corners of walls.

3) 暗：室内自然采光不好。由于只有一面小窗，而且窑室很深（常≥6m），有的窑深达33m之多，后部长年漆黑一片（砖窑、石窑可开大窗，室内光线较好）；

3) Dark: Indoor day lighting is very low because there is usually only a small window yet the cave depth is always more than 6m. In some case the depth exceeds 33m. So its rear space is always pitch dark. Brick or stone cave can have larger windows therefore their day lighting is better.

4) 占地多：窑顶自古均不种植，怕植物根系加重雨水渗透，引起塌顶。另外，由于窑脸受雨水冲塌，住户维修时，必将坍塌的窑脸土垂直铲平，再向后掘进以保持原深，年复一年，窑洞院落越来越大，相反，地面耕地却越来越少。

4) More land occupied: Since ancient times the occupants have never planted anything on top of the cave roofs. They are afraid of the plant root systems aggravating the penetration of water causing roof to collapse. Besides, driving rain often washes off the soil on cave facade. So the residents usually maintain the collapsed cave face by vertically razing it and digging further into the rear soil to keep the original cave depth. Year by year, the courtyard becomes wider and wider. As a result, the aboveground farmland becomes smaller and smaller.

5) 闭塞窑洞一般无空气循环道路，通风都不好。

5) Poor ventilation: In caves generally, there is no route for air circulation so ventilation is always bad.

(2) 综合治理

实验地区气候

实验地区（乾县张家堡）属半干旱地区，全年雨量有限，一般为400多毫米，但夏季常有连阴雨。夏湿热，室外气温高达38~42℃。冬干冷，室外气温可低至-10℃，或更低些。

该地区地下水位很深，在地表150~200m以下。乾县离西安不远，地下温度场可参考西安地区实测地温曲线（图7-19，由夏云根据我们实测17万多数据记录绘成）。

(2) Comprehensive management

Climate conditions of the experiment's site

The experiment's site (Zhangjiabu village, Qianxian County, Shaanxi Province, Northwest China) is in a semi-arid zone. It has limited annual rainfall of about 400mm. But in summer there are often continuous rainy days. Summer is warm and wet and maximum outdoor air temperatures may reach 38°~42℃. Winter is cold and dry and minimum outdoor air temperatures may be 10℃ below zero or even lower.

On the site, the water table is very deep, at least 150~200m below ground. Qianxian is not far from Xian City. So the underground temperature curves in Xian region are

7 EARTH SHELTERED BUILDINGS

taken as a reference for the underground temperature field in Qianxian. (Fig. 7-19 is drawn by Xia yun based on our measured more than 0.17 million recorded data).

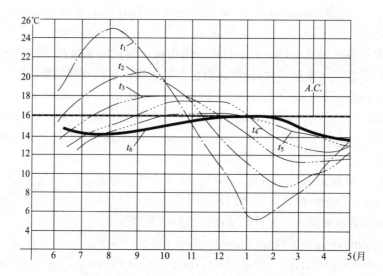

图 7-19 西安(北纬 34°15' N)地区实测地温曲线：

Fig. 7-19 Measured soil temperature curves in Xi'an (Lat. 34°15'. N.)：

$t_1 \sim t_6$：1 到 6m 深处土壤温度；$A.C$：年恒温线(15m 深处)

soil temp at 1to 6m depth respectively； $A.C$：Annual constant temp line (depth 15m)

图 7-20

Fig. 7-20

(a) 实验窑洞改建前视景；(b) 实验窑洞改建后视景

(a) Experimental cave's situation before remodeling；(b) Experimental cave's situation after remodeling

  针对前述缺点，我们提出了4条(16个汉字)改进措施：隔水防塌，育土种植，综合用能，自然空调(详见 7.3.4)。结果做到了：洞顶为田，洞中为室，生产、生活两相宜，节地又节能的效果。现分条叙述如下。

  Focusing on the shortcomings discussed above we suggested four improvements, in 16 Chinese words: excluding water to prevent collapse; backfilling soil on top of waterproof-

7.3 中国窑洞
7.3 Chinese Caves (Yao dong)

ing layer for planting; comprehensively using energy and natural air conditioning (detail in 7.3.4). Our results are: the cave roof is used as farmland while the inside remains as living space. This not only benefits production and living but also save energy and land. We'll discuss the results respectively as follows:

1) 为提供耕地和防止塌顶，设置水平防水层（一毡二油或一层塑料薄膜），其上回填 600mm 土作为种植层，种植层与防水层之间设 10cm 厚石作为滤水层和警报层。当耕者碰击豆石层发出响声警告，从而保护防水层。

1) To provide farmland and prevent roof collapse. We have installed a horizontal waterproofing membrane (one layer double-sided bituminous felt or one layer plastic film). On top of the waterproofing, 600mm thick soil was backfilled as a planting layer. Between the waterproofing and the planting layer we used a layer of 10cm thick pea gravel layer as a filter and a warning layer. Because it will send warnings sound when the farmers dig deep enough to touch gravels so as to protect the waterproofing membrane.

2) 窑脸采用砖贴面作为垂直防水层，防止窑脸土坍落。

实验证明，水平和垂直防水层对防止窑顶坍塌及窑脸土坍落是有效的。实验期曾遇连雨，该地区全部窑洞坍塌的约为 30%，其余多渗漏。但实验窑洞未见任何渗漏痕迹，防水层上设种植层是可行的。该户农民在窑顶种植层当季还收获了 350kg 红薯和一些烟叶、蔬菜。

2) The cave facade was covered with veneered bricks as the vertical waterproofing to protect the cave face from collapse. The experiment has proved that the horizontal and vertical waterproofing have effectively protected the cave roof and facade from collapsing. During the experimental period continuous rainy days occurred. In the area about 30% of the caves collapsed, and most other caves leaked. But in the experimental cave no leak appeared and the plantable layer on top of the waterproofing is workable. During the experimental season the owner of the cave has harvested 350kg sweet potatoes and some tobacco leaves and vegetables.

3) 采用较大的窗户改善自然采光及冬季太阳能直接得热供暖。老式黄土窑洞窗户都偏小，一般 20m² 左右的窑洞，窗户面积只有近 1m²，故室内甚暗。改建窑洞将窗户加大到 5.2m²，自然采光经实测，在洞室前、中、后部位分别提高到老式窑洞的 5～30 倍。实验窑洞后部老女主人能做针线活。

3) We have installed a larger window to improve day lighting and direct gain solar heating in winter. The window in old loess cave is usually very small. For example, the window is only about 1m² in a 20m² cave therefore inside lighting is very poor. We have increased the window area to 5.2m². According to measured data, the day lighting levels inside the cave, at front, middle and rear were improved respectively 5 to 30 times to the levels inside the old cave. At rear part of the experimental cave, an old female occupant can do needlework.

7.3.4 自然空调系统

图 7-21(a) 为具有本自然空调系统的掩土自然空调太阳房剖面，由前述下沉式窑洞改

7 掩 土 建 筑
7 EARTH SHELTERED BUILDINGS

建而成。

剖面图中，左侧显示冬季太阳能直接得热供暖，右侧为"⌐"钩形风道自然空调系统。图 7-21(b)为现场实测的夏季实验数据选摘。图 7-21(c)为室外空气经该风道入窑室冬增温夏降温典型曲线。可以看出，垂直风道"D"段是冬增温夏降温的决定性区段。

7.3.4 Natural air conditioning system (NACS)

Fig. 7-21(a) is the section of the earth sheltered solar house with the NACS which was remodeled from the loess cave described above.

The left of the section shows direct gain solar heating in winter and the right shows a "⌐" hook-shaped ventilation tunnel of the NACS. Fig. 7-21(b) shows the selected temperature curves measured on site in summer. Fig. 7-21(c) shows the typical curves of outdoor air temperatures increased in winter and decreased in summer when the air goes through the tunnel entering the cave room. We can see that the vertical tunnel "D" is the key part to increase the air temperatures in winter and decrease it in summer.

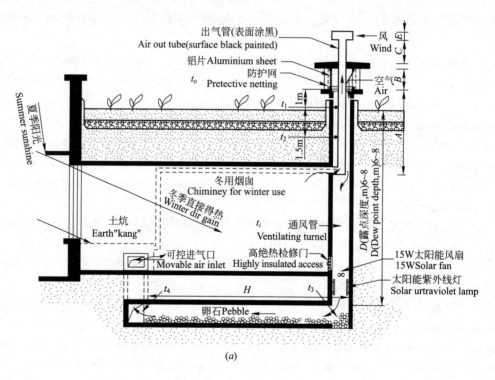

图 7-21(a) 掩土自然空调太阳房剖面

Fig. 7-21(a) The section of the earth sheltered solar house with a NACS;

t_0—室外气温 outdoor air temp; t_i—室内气温 indoor air temp;

t_1—t_4 风道内不同点气温。室外气温高达 37.1℃，室内气温仍为 22℃ temperatures at different points in the tunnel. When the outdoor temp was high of 37.1℃, the indoor air temp was still kept at 22℃

7.3 中国窑洞
7.3 Chinese Caves (Yao dong)

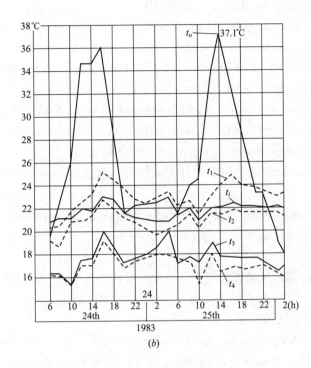
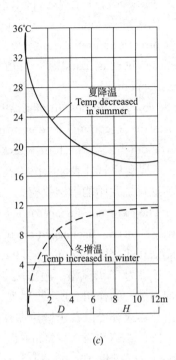

图 7-21(b)；(c)：实测温度曲线选；(c) 自然空调系统对入室空气冬增温夏降温典型曲线

Fig. 7-21(b)；(c)：Selected temperature measurement curves；(c) Typical curves of the air temperatures decreased in summer and increased in winter when the outdoor air coming into the room through the NACS；

t_o—室外气温 outdoor air temp；t_i—室内气温 indoor air temp；

t_1—t_4 风道内不同点气温。室外气温高达 37.1℃，室内气温仍为 22℃ temperatures at different points in the tunnel. When the outdoor temp was high of 37.1℃, the indoor air temp was still kept at 22℃

　　该自然空调系统究竟是怎样运行呢？细看"⌐"钩形风道，主要由垂直风道 A、B、C、D、E 构造段与水平风道 H 段组成：

　　A 段为双套管热交换器。冬季室内热脏空气经金属内套管排出，室外鲜、冷空气经砖砌风道（外套管）向下流动时，将受到出气加热。夏季，室外热、鲜空气进入风道 A 段，将受到排出的室内冷、脏空气隔污吸热的降温作用。

　　How does the NACS operate? Look carefully at the "⌐" hook-shaped tunnel which is mainly composed of the vertical tunnel parts A, B, C, D, E and horizontal part H.

　　Part "A" comprises a double-tube heat exchanger. In winter indoor hot-dirty air is discharged via the inner metal tube while outdoor cold-fresh air comes down via the brick tunnel (outer tube). The out going air will heat the incoming air. In summer, outdoor hot-fresh air, coming down via the part "A" outer tunnel, will be cooled by the out going air whose stale part will be separated from exchange.

　　B 段为捕风器。该段为砖砌方形风道，在顶端钢筋混凝土盖板下四面有进风孔，每面用薄铝片在风孔内侧装成上悬式活动遮孔板。当风吹开其中一面铝片时，风被折射向下进入垂直风道。其余三面，由于其外侧均处于负压侧，会自动关闭，使进入风孔的空气向下进入垂直风道。4个风孔外侧设有4片金属网,阻截蛇、鼠类动物入室。

Part "B" comprises a wind catcher with a square brick tunnel. There are four inlets on four directions under the top reinforced concrete cover of the tunnel. A top-hung thin aluminium sheet is installed as inlet movable cover inside each inlet. When a cover among the four is opened by wind, the wind will be deflected down into the vertical tunnel. The other three covers will be automatically shut because of the negative pressure outside them. That makes the in coming air down into the vertical tunnel. 4 pieces of metal mesh are installed on the outside of the four inlets to stop snakes and rats to come into the room.

C 段金属管外表面涂黑,在阳光下,管内被排出的空气将被加热,增加热压抽气的作用,这就是太阳能烟囱效应。

The external surface of the metal tube of part "C" is painted black. Under sunshine, it will heat the discharged air inside to increase the thermal pressure drawing air out. This is so called solar chimney effect.

E 段为 T 形风帽。根据风动力学可知,风帽外空气静压力比风管内空气静压力小,起着加强排气的作用。所以 "A" "B" "C" "E" 4 部分实际是一组综合利用太阳能、风能、室内弃热或弃冷以及土壤热惯性能的综合用能构造。

Part "E" is a "T-shaped" hood. According to the wind kinetics law, the static air pressure outside the hood tube is less than that inside the tube. This can increase the power drawing air out. So the "A", "B", "C", "E" four parts are truly a group of structure which can comprehensively use solar energy, wind energy, recycled internal abandoned heat or cold and the earth thermal inertia energy.

垂直风道 D 段区为对进气调温段(冬增温、夏降温)。例如从图 7-19 实测地温曲线可看出:最冷月是 1~2 月,土温自深 1m 至深 6m 处将从 5.5℃升至 16.5℃,故进气也将被逐渐加热,再经水平风道 H 段末端上升,由可调风孔进入室内。曾实测:当室外气温为 −6.4℃到 4.9℃时,该空气经风道入室后气温升为 8.8~12.7℃。很明显,由免费的自然能将空气加热升温 7.8~15.2℃。

Part "D" is a vertical tunnel to adjust the incoming air temperatures (increases temperatures in winter, reduces temperatures in summer). From measured soil temperature curves in Fig. 7-19 we can see, during the coldest months in Jan~Feb the soil temperatures at 1m~6m below ground are gradually raised from 5.5℃ to 16.5℃. So the incoming air will be gradually heated, then via the end of horizontal tunnel part "H" goes upwards into the room via the controllable vent. We have measured that when outdoor air temperatures were −6.4℃ to 4.9℃, air temperatures raised to 8.8℃~12.7℃ through the tunnel into room. Obviously, the incoming air was heated at 7.8℃~15.2℃ temperature rising by the natural energy without cost.

水平风道 H 段作用有二:一是调节气温、湿度,二是使入室鲜空气靠近房间前端进入室内,便于居民有一较大空间吸到新鲜空气。呼出含二氧化碳的空气则经房间后部顶端出气孔排走。H 段应处于如图 7-21(a)所示露点深度,地表下深度 6~8m 的土层在热季(6~9 月)温度均≤15℃(图 7-19)是合适的露点深度。室外高温高湿空气经该自然空调风

7.3 中国窑洞
7.3 Chinese Caves (Yao dong)

道可降温降湿,进入室内后,室温可控在22~24℃,相对湿度可控在≤70%(舒适范围)。冬季NACS对入室的干冷空气则可起到升温增湿作用。

The horizontal tunnel part "H" has two functions. One is to adjust air temperature and humidity. Another is to lead the fresh air into the room from the front part so as to allow the occupants getting a larger space to breathe in fresh air. The breathed out air containing CO_2 can be extracted out via the rear vent. Part "H" should be located at the Dew Point Depth as shown in Fig. 7-21(a). The soil layer at 6~8m below ground is an appropriate Dew Point Depth because during the hot months (Jun to Sept) its temperature is ≤15℃ (see Fig. 7-19). Outdoor air with high temperature and high humidity can reduce its temperature and humidity via the NACS and then goes into the cave space. The air temperatures can be controlled between 22~24℃ and relative humidity can be controlled around≤70% (in the comfort range). In winter the NACS can increase both the temperature and humidity of incoming cold-dry air.

经该自然空调系统可调进风孔入室的气流速度经测定,一般在0.2m/s到0.3m/s,通风道净面积可按0.5m² 每百人估算。若一户4~6人,该风道净面积考虑热、湿交换及施工要求可取0.05m²。

The measured air flow speed, via the controllable vent, going into the cave, was generally within 0.2~0.3m/s. The net ventilation area of the tunnel may be calculated at 0.5m² for each hundred people. Considering the needs of heat and humidity exchange and construction feasibility, the net ventilation area of the tunnel is 0.05m² for a family of 4~6 persons.

在风道适当处(例如风道底部)安装2盏太阳紫外线灯可杀菌、消毒;安装一座15W太阳能风扇,以便无风和热压不够时保证继续通风。

At a suitable position of the tunnel (such as at the bottom of the tunnel) two ultraviolet lamps for disinfecting (sterilizing) are installed. When there is no wind or the thermal pressure is not enough, a 15W solar powered fan is necessary to ensure the continuing ventilation.

概括起来:让湿热空气经过一段热惯性足够的露点空间进入使用房间,该入室空气便可得到调温调湿。图7-22为3种方案供参考:

(a) 图是利用深厚土层作热惯性物质对入室空气进行调温调湿。

(b) 图为利用地下水或自来水构成建筑物底层地下水池,或多层、高层建筑贮水层构成自然空调系统,并有利于防火。

(c) 图为利用充水墙体,或充水楼板构成自然空调系统。

上述方案能循环使用地下水是好方法。

In summary if hot-humid fresh air passes through a dew point space with enough thermal inertia into a habitable space, the air's temperature and humidity will be adjusted. Fig. 7-22 shows three reference schemes:

(a) is a very thick soil as the thermal inertia mass to adjust temperature and humidity of incoming air.

(b) uses available ground water or tap water to fill up a pool in the building basement

7 掩土建筑
7 EARTH SHELTERED BUILDINGS

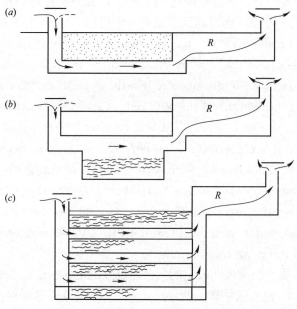

图 7-22 调温、调湿的热惯性参考方案

Fig. 7-22 Thermal inertia schemes for adjusting air temperature and humidity;
(a)深厚土层; (b)水池; (c)充水墙体或充水楼板
(a) thick soil; (b) pool; (c) water loaded wall or slab;
R—使用房间 room

or a water storage tank for multi-story/high-rise building. It becomes a NACS as well as a fire fighting facility.

(c) uses water loaded wall or slab to form a NACS.

In the above schemes it is better to recycle ground water usage.

我们设计的该自然空调系统与可种植毗连温室相结合的"太阳能水、土热惯性富氧自然空调构造"已获中国国家发明专利权。感谢医学教授张宜仁的热心帮助与指导。

This NACS combining attached plantable greenhouse is named as "Solar Energy and Oxygen-rich NACS Construction Using Water and Soil Thermal Inertia". This structure detail has been awarded the Invention Patent Right in P. R. China. Many thanks to medical Prof Zhang Yiren's warm hearted help and guidance.

7.3.5 掩土建筑卫生状况

掩土建筑内的卫生状况如何,一直是人们关注的问题。

张宜仁医学教授完成的有关医学试验与调查,如:菌落数测定,人体生理试验,室内温度比较以及人寿调查证实:掩土建筑具有有利于人类生存的条件。

7.3.5 Hygiene condition in earth sheltered houses

People have always paid close attention to the hygiene condition in earth sheltered houses. Medical professor Zhang Yiren has completed relevant medical tests and investigations

7.3 中国窑洞
7.3 Chinese Caves (Yao dong)

such as colony number test, human physiological test, indoor temperature comparisons and lifespan (life period, life time) investigation. They have proved that earth sheltered houses have conditions benefitting human existence.

(1) 菌落数测定

菌落数测定是检验所在环境细菌情况的一种医学试验。试验器是一个直径10cm的细菌培养皿，内有营养丰富培植细菌的物质。试验时将该皿置于选好的试验点暴露10分钟，将该皿编号，盖紧，携回，放入培养室培养24小时，再用显微镜点数细菌集团（菌落）数（因为单个细菌太小，难以数清）。菌落数越多，表明所测环境细菌污染越严重。

我们曾在同一城市的市中心、某大学校园、儿童公园进行了掩土空间与非掩土空间的菌落数测定。此外，我们对7.3.4所述自然空调系统也进行了同样测定。

(1) Colony number test

The colony number test is a medical test for checking germs' existence. The tester is a 10cm diameter cultivating vessel. It contains abundant nutrients for cultivating germs. During testing, the tester is exposed at the selected point for 10 minutes. Then number the tester and cover it closely. Bring all the testers back and put in the cultivating room for 24 hours. Then under microscope count the germ groups (colony number) (because a single germ is too small to number). The more colonies identified, the more tested place polluted by germs.

Colony number tests have been carried out in earth-sheltered buildings, non-earth sheltered ones, at a city center, a campus and a children park in the same city. Furthermore, we have done the same test on the NACS explained in 7.3.4.

详情如下：

1) 该市中心

在一住宅第4层露天阳台上测试（图7-23）。结果：显微镜下，菌落数多得难以数清，表明该处细菌污染严重。

The results are as follows：

1) City center

We tested at a 4th floor open balcony in a dwelling (Fig. 7-23). The result is that the colony numbers were too great (much) to count under the microscope. It indicated that the place was severely polluted by germs.

图 7-23 张教授在某市中心一阳台上测试菌落数

Fig. 7-23 Prof Zhang was doing the colony test at a balcony in a city center

2) 儿童公园（绿化一般）（图7-24）

室外菌落数为40，低水平洁净。而其掩土地下厅（RC拱形结构，最薄处覆土厚2m）菌落数只有2个，最洁净。

2) Children's park (ordinary landscaping) (Fig. 7-24)

Outdoor colony number is 40, low hygiene. In the earth-sheltered hall (RC arch, the thinnest cover soil is 2m), the colony number is only 2 and it's the cleanest.

3) 某大学校园（绿化较好）（图7-25）

7 掩土建筑
7 EARTH SHELTERED BUILDINGS

室外菌落数为 28，紧邻的一个四层办公室(砖混结构)，菌落数为 9，洁净，而紧邻的一地面掩土建筑(图 7-25)内菌落数只有 2 个，最洁净。

3) A university campus. (Better landscaping) (Fig. 7-25)

Outdoor colony number is 28, colony number is 9 at an adjacent 4-story office building (concrete-brick construction), it is clean. Yet in an adjacent aboveground earth sheltered building (Fig. 7-25), the colony number is only 2 and it is the cleanest.

图 7-24 某市儿童公园掩土地下厅
Fig. 7-24 An earth sheltered hall in a city children park

图 7-25 某大学地面掩土建筑
Fig. 7-25 An aboveground earth sheltered building in an university

4) 自然空调系统菌落数

当室外菌落为 46(细菌环境坏)时，该自然空调系统垂直进气管道底部菌落数仅为 4，很洁净。这表明该自然空调系统不仅有调节温、湿度的功能还有沉淀细菌净化空气的功能。安装两套太阳能紫外线灯，将得到更好的灭菌效果。

4) Colony number in the NACS

When outdoor colony number was 46(bad germ environment), at the bottom of the vertical tunnel of theNACS the colony number was 4, very clean. It indicates that the NACS not only has the ability to adjust the air temperature and humidity but also has the capacity to purify the air by sedimentating the germs. We'll get better result of killing germs if two ultraviolet lamps are installed.

(2) 人体生理试验

为比较在掩土建筑与地面传统建筑内人体的生理反应，进行了冬、夏两次人体生理反应试验。

冬季人体生理试验在陕西乾县张家堡改建的下沉式窑洞与邻近一小学教室进行。实验对象为 5 男 5 女 12 到 14 岁的小学生(图 7-26)(选择少年学生为试验对象是因为他(她)们受其他疾病因素影响小)。试验表明，他(她)们绝大多数(70%)逗留在教室时脉搏较逗留在该掩土建筑内每分钟要快 2~10 次，呼吸每分钟也快 2~4 次；相反，他(她)们的鼻端温度(按中医理论，鼻端乃热敏感穴位)全都低于逗留在掩土建筑内各自的鼻端温度(低 2.8~10.3℃)。当时教室内温度只有 6℃，掩土建筑内室温 11.5℃，两者均无供暖，室外气温 1.2℃。

7.3 中 国 窑 洞

7.3 Chinese Caves (Yao dong)

(2) Human physiological test

To compare human body physiological response living in earth sheltered buildings with living in aboveground conventional buildings, we had done two tests, one in winter, another in summer.

The winter test was done in a remodeled sunken cave and a nearby primary school classroom, at Zhangjiabu village, Qianxian County, Shaanxi Province. Participants to be tested are ten pupils of 5 boys and 5 girls aged between 12~14. (Fig. 7-26) (Pupils are less affected by other illnesses so we select them to be tested). The test indicated that most (70%) of their pulse times/m in the classroom was 2 to 10 times/m faster than tested in the earth sheltered building. Their breath times/m were also 2 to 4 times/m faster. Oppositely, their nose tip temperatures (according to traditional Chinese medicine theory nose tip is a thermal sensitive point) were all 1.8 to 10.3℃ lower than that of them staying in the earth sheltered building. The classroom air temperature was then 6℃, while the earth sheltered building's room temperature was 11.5℃. Both were no heated and the outdoor temperature was 1.2℃.

夏季试验是在西安市一大学校园内的掩土建筑及其邻近单层办公室内进行的。试验对象为中学生5男5女(图7-27,女生未入图)。结果表明,70%的人在单层办公室内的每分钟脉搏数比在掩土建筑内快2~14次,80%的人每分钟呼吸次数快1~10次,他(她)们的鼻端温度全都比在掩土建筑内鼻端温度高。当时,单层办公室当时的室温33.1℃,掩土建筑内室温29.2℃。

The summer test was done in an earth sheltered building and a nearby single storied office building in the campus of a university in Xi'an City. 5 boys and 5 girls from middle school were tested (Fig. 7-27, no girl students in the picture). The result showed that the pulse times/m of 70% students staying in the office were 2 to 14 times/m faster than that of those staying in the earth sheltered building. 80% of their breath times/m were 1 to 10 times/m faster than that of those staying in the earth-sheltered building. Their nose tip temperatures were all higher than that of those staying in the earth sheltered building. The single storied office room temperature was then 33.1℃, while the earth sheltered building room temperature was 29.2℃.

图 7-26 冬季人体生理试验

Fig. 7-26 Physiological test in winter

图 7-27 夏季人体生理试验

Fig. 7-27 Physiological test in summer

7 掩 土 建 筑
7 EARTH SHELTERED BUILDINGS

上述试验充分证明，无论是冬季或夏季，掩土建筑不仅能节约传统能源保持室内较佳微气候，还能节约人体新陈代谢，和改善生理状况，这确是长寿因素。（张宜仁教授的新发现）

试验详细数据见表 7-1a（汉语表），7-1b（英语表）

The above tests have sufficiently proved that whether in winter or summer the earth sheltered building not only can save conventional energy to keep better indoor microclimate but also can save the human body's metabolic energy while improving physiological conditions. This is truly a longevity factor.

The tested detail data was shown in table 7-1(a) (in Chinese) and Table 7-1(b) (in English).

掩土建筑与传统建筑内人体生理试验比较　　　　　　　　表 7-1(a)

受试少年学生	年龄	冬季试验 脉搏(次/分) 在下沉式掩土建筑内	在地面教室内	呼吸(次/分) 在下沉式掩土建筑内	在地面教室内	鼻端温度℃ 在下沉式掩土建筑内	在地面教室内	受试少年学生	年龄	夏季试验 脉搏(次/分) 在下沉式掩土建筑内	在地面教室内	呼吸(次/分) 在下沉式掩土建筑内	在地面教室内	鼻端温度℃ 在下沉式掩土建筑内	在地面教室内
No.1 女	12	72	74	24	26	18.2	14.0	No.1 女	13	70	84	17	20	31.4	33.7
No.2 女	13	80	84	18	22	22.0	15.5	No.2 女	13	100	100	19	20	32.4	34.3
No.3 女	13	110	100	24	24	24.0	17.2	No.3 女	13	120	120	20	22	33.6	35.0
No.4 女	13	78	80	18	18	18.0	15.2	No.4 女	14	100	110	19	18	33.4	34.5
No.5 女	13	90	100	20	18	21.5	17.2	No.5 女	14	78	66	16	16	33.3	34.3
No.6 男	14	60	60	20	22	16.5	13.5	No.6 男	12	82	93	16	26	33.1	33.3
No.7 男	14	80	84	16	18	26.5	17.0	No.7 男	13	80	82	17	18	32.5	33.7
No.8 男	14	90	100	20	22	21.5	17.0	No.8 男	13	77	100	17	21	33.5	33.7
No.9 男	14	60	60	18	20	23.5	18.5	No.9 男	13	72	80	17	18	33.3	33.9
No.10 男	14	72	76	18	20	25.5	15.2	No.10 男	14	80	84	18	22	33.2	34.4

在下沉式掩土建筑内气温 11.5℃
地面教室内气温 6℃

地面掩土建筑内气温 29.2℃
地面办公室内气温 33.1℃

physiologic test comparison between in earth sheltered building and conventional building　　Table 7-1(b)

Pupils tested	Age	Winter test Pulse (Times/m) In earth sheltered building	In Class room	Breath (Times/m) In earth sheltered building	In Class room	Nose tip temp, (c) In earth sheltered building	In Class room	Pupils tested	Age	Summer test Pulse (Times/m) In earth sheltered building	In Class room	Breath (Times/m) In earth sheltered building	In Class room	Nose tip temp, (c) In earth sheltered building	In Class room
No.1 girl	12	72	74	24	26	18.2	14.0	No.1 girl	13	70	84	17	20	31.4	33.7
No.2 girl	13	80	84	18	22	22.0	15.5	No.2 girl	13	100	100	19	20	32.4	34.3
No.3 girl	13	110	100	24	24	24.0	17.2	No.3 girl	13	120	120	20	22	33.6	35.0

7.3 中 国 窑 洞
7.3 Chinese Caves (Yao dong)

Continued

Pupils tested	Age	Winter test					Pupils tested	Age	Summer test				
		Pulse (Times/m)		Breath (Times/m)		Nose tip temp, (c)			Pulse (Times/m)		Breath (Times/m)		Nose tip temp, (c)
		In earth sheltered building	In Class room	In earth sheltered building	In Class room	In earth sheltered building	In Class room		In earth sheltered building	In Class room	In earth sheltered building	In Class room	In earth sheltered building	In Class room
No. 4girl	13	78	80	18	18	18.0	15.2	No. 4girl	14	100	110	19	18	33.4	34.5
No. 5girl	13	90	100	20	20	21.5	17.2	No. 5girl	14	78	66	16	16	33.3	34.3
No. 6boy	14	60	60	20	22	16.5	13.5	No. 6boy	12	82	93	16	26	33.1	33.3
No. 7boy	14	80	84	16	20	26.5	17.0	No. 7boy	13	80	82	17	18	32.5	33.7
No. 8boy	14	90	100	20	22	21.5	17.0	No. 8boy	13	77	100	17	21	33.5	33.7
No. 9boy	14	60	60	16	18	23.5	18.5	No. 9boy	13	72	80	17	18	33.3	33.9
No. 10boy	14	72	76	18	20	25.5	15.2	No. 10boy	14	80	84	18	22	33.2	34.4

In earth sheltered building air temp11.5℃
In classroom air temp6℃

In earth sheltered building air temp 29.2℃
In classroom air temp33.1℃

(3) 室内温度波动比较

图 7-28 示出地面传统农房 A)、地面掩土房 B)、与下沉式掩土房 C)室内温度年波动比较。从图中清楚地看出，下沉式掩土建筑室内温度最稳定，最高、最低年波动温差仅 11℃(11～22℃)；地面掩土建筑次之(6.5～25℃)，年高低温差 18.5℃；地面传统农房室内温度稳定性最差，年高低温度差达 32.1℃(－0.1～32℃)。图 7-28 下图所示土层年温度波动随深度作喇叭形衰减，表明了上述不同建筑内温度波动差异的缘由。

(3) Indoor temperature fluctuation comparison

Fig. 7-28 shows the annual temperature fluctuation comparison among the conventional aboveground rural house, aboveground earth sheltered house and the sunken earth sheltered house. From the figure we can clearly see that in the sunken earth sheltered house the room temperature was most stable. The annual temperature difference between the highest room temperature and the lowest one was 11℃(11～22℃). In the aboveground earth sheltered house, the room temperature was second stable with annual fluctuation of 18.5℃ between the highest and lowest (6.5～25℃). In the conventional above ground rural house, the room temperature stability was the worst with the annual fluctuation reaching 32.1℃(－0.1～32℃). The lower diagram of Fig. 7-28 shows the soil temperature annual fluctuation attenuates with the soil depth and appears in the shape of a trumpet. This shows the reason for the differences of temperature fluctuations in the different houses shown above.

(4) 掩土建筑与健康、安全

以上三医学试验：菌落试验、人体生理试验以及温度波动比较都证明了掩土建筑对健康的有利条件。此外，掩土建筑有良好的功能：防震(含地震)、防风、防尘暴、隔噪声、

7 掩土建筑
7 EARTH SHELTERED BUILDINGS

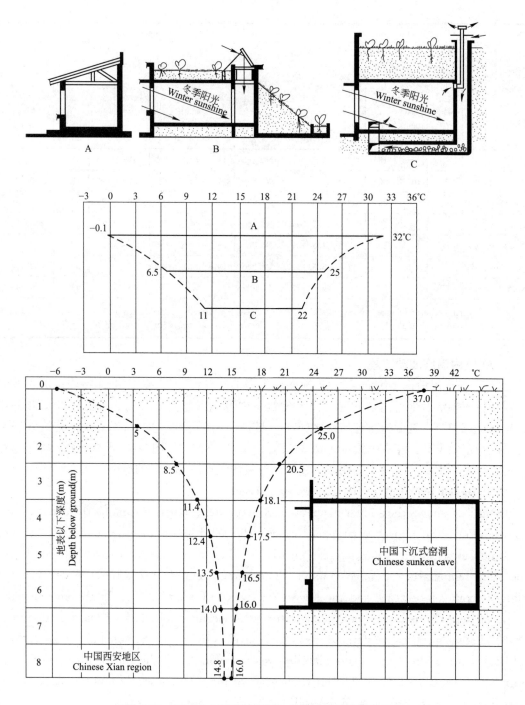

图 7-28 不同室内年温波比较与土层温波随深度的衰减

Fig. 7-28 The comparison of indoor annual temperature fluctuations in different rooms and soil temperature fluctuations attenuated down with soil depth

防止放射性污染及大气污染侵入，阻止火灾蔓延，比非掩土建筑更安全，生态平衡条件好，全都是有利于健康、安全的因素。

7.3 中国窑洞
7.3 Chinese Caves (Yao dong)

(4) Earth sheltered building related to health and safety

The above three medical tests: colony test, physiological test and indoor temperature fluctuation comparison have all proved that the earth sheltered building has conditions benefiting health. Besides, the earth sheltered building has strong capacity in against quakes, wind, dust storm, noise, radiant pollution, air pollution and fire spreading. Compared with non-eat sheltered buildings, earth sheltered buildings are safer and have better conditions of eco-balance. All of the above are factors benefiting health and safety.

在对住窑洞的居民人寿调查中，发现长寿的人确实不少(特别是女性)。图 7-29 所示四位长寿女老人住在河南巩县康南村(十几户人家)，都是从小一直住窑洞，最长寿的一位(倚立于门边)97 岁，其余三位分别为 84、73、64 岁(1989 年)。

图 7-30 为又一位延安一生住窑洞的长寿老人。86 岁(1989)，他仍一直领导他的后代为人建石窑。

From the investigation of life-span of people living in caves, we have found more aged people (especially women). Fig. 7-29 shows 4 very old women living at Kangnan Village (more than ten families) Gong County, Henan Province. They have been living in caves from childhood. The oldest one (against the gate) was 97 years old. The other three aged 84, 73, 64 (in 1989) respectively.

Fig. 7-30 shows another long-lived old man, aged 86 (in 1989), who has spent his entire life living in caves. He has been leading his families (offspring) to build stone caves for others.

图 7-29 河南巩县康南村住窑洞长寿老人
Fig. 7-29 Long-life women living in caves, Kangnan Village, Gong County,
自左至右：97 岁、84 岁、73 岁、64 岁(1989 年)
Henan Province. Left to right, ages:
97, 84, 73, 64 (in 1989)

图 7-30 一位长寿老人，(86 岁)一生住窑洞、建窑洞(1989 年 摄影)
Fig. 7-30 A long-life man (age: 86): living in caves in his entire life and building caves for others (photo: 1989)

(5) 防止氡污染

氡是一种强致癌气体，无色、无味、无臭。它是由铀、钍放射性元素衰变而来(见下表 7-2 解)。地基土是最主要的源地。

防止室内氡污染建筑法：
① 通风（自然空调通风最好）。
② 内表面紧密抹灰，面与面交接处做弧形接缝。

(5) Stopping radon pollution

Radon is a strong carcinogen radiant gas without color, taste or smell. It comes from the decay course of the radiant elements: uranium and thorium (see the following diagram). Subsoil is its main source. Architectural methods stopping indoor radon pollution are:

① Ventilation (NAC is the best).

② Inner surface tightly rendered, coved joints between faces.

铀、钍放射性元素衰变　　　　　　　　　　　　　　　　表7-2
The decay couse of uranium and thorium　　　　　　　Table 7-2

239 u→	235 Ra→	222 Rn→	218 PO→	214 Pb→	214 Bi→	205 Pb
232 Th→	224 Ra→	220 Rn→	216 PO→	212 Pb→	214 Bi→	208 Pb
235 u→	223 Ra→	219 Rn→	215 PO→	211 Pb→	211 Bi→	207 Pb
铀 uranium 钍 thorium	镭 radium	氡 radon	钋 Polonium	铅 Lead	铋 bismuth	铅 Lead

7.4 中国城市地下空间

7.4.1 城市地下空间概况

中国内地城市地下空间现代技术的开发利用只是国家改革开放以后才有了较好的起步。北京、上海、广州、深圳均已建成地铁并投入使用，西安等城市地铁正在建造中（2011年主要路段已开始营运）。此外，城市地下商场、影院、旅馆等在不少城市也有了发展。但是，与发达国家相比，我们的差距仍很大。例如匈牙利首都布达佩斯2百多万人口，就有红、黄、蓝三层地铁。

7.4 Underground Spaces in Chinese Cities

7.4.1 General picture of underground spaces in cities

In Chinese mainland the development and utilization of underground spaces by modern technique have got a better start only since the reform and opening of the country. Beijing, Shanghai, Guangzhou, Shenzhen have completed their subway systems and in operations. Xi'an and some other cities' subways are under construction (Since 2011, the main parts have doen business). Furthermore, some cities have developed shops, theatres, hotels, and etc in underground spaces. But compared with developed countries we still fall far behind. For example, in Buda-

7.4 中国城市地下空间
7.4 Underground Spaces in Chinese Cities

pest, the capital of Hungary, the population is a little more than two millions but there are three levels' subways colored "Red", "Yellow" and "Blue".

英、法、美、德、匈、瑞典、挪威、瑞士、日本、加拿大等国人口加起来还不到我国人口的一半。但从第二次世界大战结束后,他们在重建和扩建地面建筑的同时大力开发现代地下空间。半个多世纪以来,他们才有了现今巨大的成就。如前所述,在地下空间有多层地铁、停车场、火车站、发电厂、信息中心、缩微贮存库、影剧院、学校(含大学)、图书馆、精密实验室、数据处理中心、微波站、医院、住宅等。这些国家人口并不多,这么早就开发地下空间确是来自规划师、建筑师、工程师、决策者们的预见能力。前苏联在那样地广人稀的条件下也集中了20多所大学规划莫斯科多功能地下空间。莫斯科每天可运800万人次的地铁正发挥着巨大的缓解地面交通与人群的作用。日本在二战后很快就抓紧教育,积累了人才和技术,从而迅速增强了综合国力,使他们在发展现代地下空间居领先地位。在此领域我们有差距,但随着综合国力的不断增强,人才的不断增长,越来越多的规划师、建筑师、工程师、决策者在处理我国人口、能源、土地、建房、环境的错综矛盾中已越来越认识到城市,特别是大城市,开发利用现代地下空间的必要性、迫切性和可能性。

The total population of UK, France, USA, Germany, Hungary, Sweden, Norway, Switzerland, Canada, and Japan is less than half of the population of China. But since the end of World War II in 1940 they have greatly developed modern underground spaces while reconstructed and expanded aboveground buildings. For more than half a century they have achieved excellent results. As mentioned before, in underground spaces there are multilevel subways, parking stations, railway stations, power stations, information centers, microfilm stores, theatres, schools (including universities), libraries, precision test laboratories, data centers, microwave station, hospitals and dwellings. These countries' population is not vast but they have developed underground spaces so early just because of the foresight of the planners, architects, engineers and policymakers. In the former CCCP with such a vast territory but low density of population, there is still a scheme to united more than 20 universities to plan the multi-function underground space in Moscow. The Moscow subways with the capacity of carrying eight millions person-trips/day, have succeeded in relieving ground from traffic and crowds. After World War II Japan quickly improved education, accumulated talents and technologies so that it rapidly strengthened comprehensive capacity which puts them at the leading position in developing modern underground space. In this field we have been falling behind. But following the increasing of national comprehensive capacity and human resources and talents, when dealing with the complex contradictions of population, energy, land, housing and environment, more and more planners, architects, engineers and policymakers have realized the necessity, urgency and possibility of developing and using modern underground spaces in cities especially in big cities.

7.4.2 城市地下空间—范例

西安钟鼓楼广场的地下空间已成为中国现代地下空间的一个范例。彩图7-31(a)、(b)

即该广场部分视景。广场下为一大型购物中心。广场种植花草，配置石凳，还建有音乐喷泉。这座购物中心的屋顶花园已成为该市居民喜爱的活动场所之一。

（设计者：张锦秋大师及其团队）

7.4.2 A good example of underground space in city

Xian Bell-Drum Tower Square's underground space has become a good modern example of Chinese underground spaces. Color pictures 7-31((a).(b)) show a partial view of the square. There is a big shopping center under the square. In the square, flowers and grasses have been planted. Stone stools and a music fountain are installed. The roof garden of the shopping center has become one of the citizen's favorite places.

(Designer: master architect: Zhang Jinqiu and her team)

7.4.3 小结

中国的窑洞，为中华民族的延续发展已经并仍在作出巨大的贡献。祖先在下沉式窑洞中给我们留下了开发利用浅层地下空间的经验，例如冬暖夏凉、自然采光、下沉式庭院等。其中节能节地和综合用能的潜在优势正待我们以现代化科技发扬光大；靠山窑洞因山就势、就地取材、视景宽、山风好、阳光美等。这些祖先相传的宝贵经验，我们在各种坡地上建房时应大力发扬光大。

7.4.3 Summary

Chinese caves have been contributing a great deal for continuing development of the Chinese nation. With the sunken caves our ancestors left us the experiences of developing and using shallow underground spaces such as keeping warm in winter and cool in summer, getting natural lighting and sunken courtyards. The potential advantages in energy saving, land saving and comprehensively using energy are waiting for us to carry them forward by modern science and technology. In hillside caves, we can learn the experiences of locating caves according to hills' geography, using local materials, getting broad prospect and breeze and receiving nice sunshine. These precious experiences left us by our ancestors should be carry forward when we construct buildings on various hillside lands.

历代王朝的陵墓以及宗教界的地下庙殿为我们留下了：上下结合，立体用地，风水观选址和精湛的建造技术等宝贵经验。北京明十三陵、西安秦兵马俑、女皇武则天墓、唐地下佛殿法门寺等。这些掩土建筑显示的奇迹般的建筑文脉，在现代地下空间中，我们责无旁贷应加以发扬光大。

The emperors' tombs and the religious underground temples built in the past have left us valuable experiences, which include combining aboveground & underground spaces, stereoscopically using land, site selection with Fengshui theory and excellent craftsmanship and etc. The tombs of 13 Ming Dynasties emperors, Beijing, Qin terracotta figures of warriors and horses buried with the dead, Xi'an, the empress Wuzatian's tomb, the Tang dynasty underground Buddhist temple-Famen Temple, and so on, are all earth shel-

7.4 中国城市地下空间
7.4 Underground Spaces in Chinese Cities

tered buildings. They have demonstrated a wonderful architecture context. We should be responsible to carry them forward in modern underground spaces.

我国城乡开发利用地下空间的潜力极其巨大，可争得几百亿平方米的地下建筑面积。

21世纪，地下空间必将以更大的规模、更高的科技手段构成人类可持续生存空间的组成部分。

可控热核聚变、常温核聚变、超导输电、地下沼气网等因素为人造太阳、人为可持续能源以及物流的良性循环，以及能流形式的良性转换创造了条件，为建立地下绿色世界铺垫了基础。

In Chinese cities and villages there exists great potential in developing and using underground spaces. We might get several tens of billion m^2 underground area.

During the 21st century underground spaces will certainly form a component of human sustainable living spaces and be developed on a large scale by advanced science-technology.

All factors such as the controllable thermonuclear fusion, room temperature nuclear fusion, electricity transmitted by superconductor, subterranean methane web, will create the conditions for man-made sun, man-made sustainable energy and the good circulation of mass flow and the good conversion of energy forms which will construct a good basis to build the underground green world.

8 建筑绿化
8　GREENING OF BUILDINGS

8.1　光合作用的伟大贡献

光合作用：

$$6CO_2 + 6H_2O \xrightarrow[\text{绿色植物}]{\text{太阳光能}} C_6H_{12}O_6 + 6O_2$$

绿色植物通过上述光合作用将无机物二氧化碳和水变成有机物—碳水化合物，并产出氧气，其伟大贡献：

（1）为人类、动物及植物本身提供生存、发展的物质资源与能量资源。无植物也就无人类和其他动物。

（2）绿色世界是平衡大气温室效应、消除气候过度变暖及避免两极冰山消融最强大的绿色力量。

两极冰山的白色力量与植物的绿色力量是维持地球生态平衡最伟大的两位守护神！

（3）绿色植物产生的氧气（O_2）除供人类、动物、植物所需外还为臭氧层（O_3）补充原料（O_2）。

（4）绿色世界能为人类精神生活提供无限源泉。

建筑侵占绿色面积也就是危害我们自己。

建筑绿化是补偿建筑侵占绿色面积的最好对策。

8.1　The Great Contributions of Photosynthesis

Photosynthesis：

$$6CO_2 + 6H_2O \xrightarrow[\text{Greenplants}]{\text{Solarlight}} C_6H_{12}O_6 + 6O_2$$

The green plants through photosynthesis make the inorganic substance carbon dioxide and water into organic substance—carbohydrate and produce oxygen. Its' great contributions：

(1) Supply matter resources and energy resources for the existence and development of human race, animals and plant themselves. No plants no humanity and other animals.

(2) Green world is the strongest green force to balance the atmospheric greenhouse effect, eliminate climate over warming and avoid the two poles' icebergs to be melted.

The white force of the two poles' icebergs and the green force of plants are the two greatest guardian angels to keep the earth eco-balance.

(3) Green plants produce oxygen (O_2) to satisfy the needs of human beings, animals and plants also compensate the ozonosphere (O_3) material (O_2).

(4) The green world can supply a no-end headspring for human cultural life.

Buildings encroach upon green area also damage ourselves.

The greening of buildings is the best countermeasure to compensate the green area occupied by buildings.

8.2 建筑绿化的含义·效益·实例·实验

8.2.1 含义·效益·实例

建筑绿化就是尽量利用屋面、墙面、窗台、阳台、走廊、散水、水平和垂直遮阳构件、共享空间、边角可宜之处以及室外周边环境种植花草、树木、水果、蔬菜、花生、豆类以及其他园艺或农艺作物。建筑绿化搞好了，至少可得到如下生态效益：

（1）绿色面积少损失、不损失甚至增多。

建筑绿化好了可做到地基占去的绿化面积用屋面种植来补偿，道路占去的绿化面积用墙面种植来补偿。那么，绿化面积就可能少减、不减甚至增多。

（2）美化、香化、富氧环境与立体绿色园地；增多农艺、园艺产品。

（3）大城市的多、高层建筑群将不再是破坏绿化、破坏生态平衡、产生热岛负效应的巨魔，而是城市立体绿化园地的主体。

（4）广大乡村、城镇将变成星罗棋布的别墅。

（5）减轻或消除城市热岛负效应。

（6）一系列其他生态效益：防风、防尘、减噪声、吸收有害气体、调节温湿度、减轻或消除水土流失（尤其是在各种山地）。

（7）变非可持续构件为可持续构件。

8.2 The Meaning · Benefits · Examples · Experiments of Greening of Buildings

8.2.1 Meaning · benefits · examples

Greening of buildings means that as possible as we can use roofs, walls, windowsills, balconies, corridors, aprons, horizontal and vertical sunshading boards, communal spaces, available corners and sides and outdoor surroundings to plant flowers, grasses, trees, fruits, vegetables, peanuts, legumes (legumens) and other horticultural or agricultural plants. If the greening of buildings is done well, at least we can get the following eco-benefits:

(1) Green area less loss, no loss even increase.

When greening of buildings is successful, the green area occupied by foundations we can get compensation with roof planting, green area occupied by roads we can get compensation with wall planting, if so, the green area can be less decreasing, no decreasing even increasing.

(2) Environments of beautification, sweet smell, oxygen-rich and stereoscopic green gardens; increasing agricultural and horticultural products.

(3) In big cities the multi-story and high-rise buildings (highrises) will be no more the huge devils of destroying greening, damaging eco-balance and causing thermal island nega-

8 建筑绿化
8 GREENING OF BUILDINGS

tive effect, they'll become the main body of city's stereoscopic green gardens.

(4) The vast villages and towns will become gardens and villas spreading all over the country.

(5) Reducing or eliminating urban thermal island negative effect.

(6) A series of other eco-benefits: Windproofing, dustproofing, lowering noise, absorbing harmful gases, regulating temperatures and humidities, reducing or eliminating soil-erosion (especially in various mountain lands).

(7) Making non-sustainable members sustainable members.

实例：

图 8-1 是四川成都某工厂实验的屋顶种植。

Examples:

Fig. 8-1 shows an experimental roof planting at a factory, Chengdu, Sichuan.

图 8-1 某屋顶种植（广士奎教授提供）

Fig. 8-1 A roof planting (Source: Prof Guang Shikui)

试种作物每亩产量：红苕 5000kg、西红柿 6000kg、花生 200kg、黄瓜 4400kg、油菜子 150kg（1 亩＝1/15 公顷，1mu＝1/15 hectare）. Tested crops' yield per-mu: sweet potato 5000kg, tomato 6000kg, peanut 200kg, cucumber 4400kg, rapeseed 150kg.

彩图 8-2 (a)、(b) 为华南广东佛山市环境绿化两景。佛山市是我国城市绿化卓有成效的城市之一。彩图 8-2 (c)、(d) 为深圳市绿化两景。该市绿化覆盖率已达 45％。有的建筑底层做成室内花园，与室外庭园相通，组成色彩丰富、内外结合的庭园。有的高层建筑的各层凹廊组成的小花园，花、草、水、鱼、鸟都有，看上去就是一道绚丽多彩的竖向画廊。该市已被有关国际组织授予 100 万人以上的"花园城市"称号。彩图 8-2 (e)、(f) 为西安市新城广场环境绿化两景。每日清晨与黄昏人群聚此锻炼：盘刀、舞剑、弄枪、打太极拳、跳舞、溜旱冰、放风筝等。

Color pictures 8-2 (a), (b) are showing two environmental greening views of Foshan City, Guangdong Province, South China. This city is one of the Chinese cities of greening efficiency. Color pictures 8-2 (c), (d) show two greening views in Shenzhen City. The cover rate of greening of the city has reached 45％. Some buildings' first floors are planted to form indoor gardens connecting outdoor gardens to construct colourful gardens of inside combining outside. In some highrises each floor's recessed verandas are formed small gar-

8.2 建筑绿化的含义·效益·实例·实验
8.2 The Meaning · Benefits · Examples · Experiments of Greening of Buildings

dens with flowers, grass, water, fishes, birds, etc. which look bright and colourful vertical galleries. This city has been awarded the honour title of "Garden City" of population over one million by an international organization concerned. Color pictures 8-2 (e), (f) show two greening views of Xincheng Square in Xi'an City. Every early morning and evening people gather here to do exercises: playing knifes, dancing with swords, handling spears, doing Taijiquan, dancing, concrete skating with roller skates, and flying kites, etc.

建筑大师熊明在其《城市设计学》中写道:"水系不仅是城市的生命线,而且是城市的重要风景线"。以下诸例均为水—绿结合实景:彩图 8-3(a)、(b)显示西安市护城河成了环绕西安市区的一道水—绿结合的风景带、微气候调节圈。

Mr Xiong Ming, a Chinese great architect in his book《CITY DESIGN STUDY》says: "Water system is not only city's lifeline but also the city important landscape line". The following pictures are examples of combinative water-green landscapes: Color pictures 8-3 (a), (b) are showing that the Xi'an City moat has become a combinative water-green landscape belt and a microclimate regulator around the city.

彩图 8-4(a)至(e)为法国凡尔赛公园(古凡尔赛宫)水—绿结合数景:(a)图为正视景之一;(b)图与(c)图为以男女青铜雕像为主的水—绿结合一景;(d)图为音乐喷泉与周边水—绿视景;(e)图:好宽的草坪,好远的湖景,好深的树林!

Color pictures 8-4 (a) to (e) show some water-green landscapes of France Versailles Park (ancient Versailles Palace): (a), a front view; (b) and (c), male and female bronze statues as main body landscapes of combining water with green; (d), a view of the music-fountain and the surrounding combinative landscape of water and green; (e) shows a view: What a wide lawn! How a far lake sights! So deep the forest!

彩图 8-5 是法国巴黎另一世界著名风景点——埃菲尔铁塔风景区。这是一个经过精心设计和建造的以埃菲尔铁塔为标志的水—绿结合的风景花园。埃菲尔铁塔四角支承构架内有电梯上下,供游客观赏巴黎市景。

Color picture 8-5 is another world famous landscape-the Eiffel Tower landscape area in Paris, France. This is a water-green combined scenic garden with the Eiffel Tower as the mark completed by meticulous design and construction. In the four structural supporting frames at the four corners of the Tower there have lifts up/down to let travellers enjoy the sights of Paris City.

彩图 8-6 是荷兰阿伯多伦市 HETLOO 公园中世界著名的水—绿—雕像结合的美景:白玉的妇婴雕,洁白细腻,线条优美,姿态充满母爱与稚真,与环绕底部的金色粗犷健实的护神群形成强烈鲜明的对照,背景为浓绿的树林,更加衬托出雕像的优美(法国波尔多建筑学院院长 AUZANNEUA 教授提供)。

Color picture 8-6 is showing a world famous water-green-statue combinative beauty at HETLOO Park in Arbedolan City, the Netherlands (Holland), it shows a beautiful statue of a mother with her baby in white jade, showing pure white, fine smooth and graceful lines. Their posture shows mother's full love and baby's pure innocence. Around the base the guard gods with golden rough vigorous lines form a striking (sharp) contrast with the

8 建筑绿化
8 GREENING OF BUILDINGS

white statue of mother with baby. The statue's background is a deep dark green forest, it further sets off the statue more elegance (Photo source: France Bordeaux Architecture school Head, Prof. AUZANNEUA).

彩图 8-7 是新疆石河子市绿化一景。该地区年降雨量甚少（仅 199mm），但地下水丰富，主要是天山冰雪融化渗流水，有"坎儿井"地下供水系统。该系统可保持水质清洁，并可避免蒸发损失。"石河子并非满河石头子，万里蓝天遍地树，有天山固体水库，坎儿井沥水清流，碧水湖波翻白浪，欢快鱼儿跃银光，好一片绿洲胜地！"

Color picture 8-7 is showing a greening view of Shihezi City, Xinjiang region, Northwest China, here annual rainfall is very low (only 199mm), but rich of groundwater coming mainly from the permeated water originated from the melted snow and ice in Tianshan Mountains. The groundwater supply system "Kanerjing" can keep the water clean and avoid the evaporative loss. "Shihezi is not full of stones in rivers. Blue sky covers a boundless space, everywhere are growing green trees. The Tianshan Mountains are the solid reservoir. Kanerjing flowing water is always clear and clean. Dark green lake waves white waves and jumping fishes reflect silver lights. What a wonderful oasis here is!"

图 8-8 为西安市人民剧院西晒墙面有无绿化的某夏季测温比较，最大温差为 14℃（下午 5 时 10 分），有绿叶覆盖的灰墙面为 32℃，裸露面为 46℃。冬季，叶片脱落，但枝藤仍吸附在墙面，可减少对流失热。(图 8-1)为屋面绿化例，更有利于冬、夏绝热。彩图 8-9 为西安市人民剧院西墙面绿化实景。

Fig. 8-8 shows a comparison of the temperatures measured in a summer between the exposed wall and greening covered one from the west-facing wall of Xi'an People Theatre. The maximum temperature difference was 14℃(14K), the temperature of gray brick wall covered by green leaves was 32℃, of the exposed one was 46℃. In winter, leaves fall off but the branches can be still glued on the wall to decrease the convective heat loss. Fig. 8-1, an example of roof planting more benefits winter and summer insulation. Color picture 8-9 shows a view of the greening of the west-facing wall of the Xi'an People Theatre.

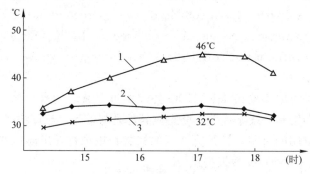

图 8-8 西安市人民剧院西墙面西晒温度比较

Fig. 8-8 A comparison of temperatures of the west-facing wall of the Xi'an People Theatre under sunshine

1—西向暴露墙面温度 Temperatures of the exposed west-facing wall; 2—室外气温 Outdoor air temperatures; 3—绿叶覆盖西向墙面温度（李莉萍提供）Temperatures of the greening covered west-facing wall. (Source: Liliping)

8.2 建筑绿化的含义·效益·实例·实验
8.2 The Meaning · Benefits · Examples · Experiments of Greening of Buildings

8.2.2 阳台冬季温室效应实验

例：西安某向阳阳台，安装单层玻璃推拉窗内侧悬一层透明农用塑料薄膜，在窗顶保留高5cm，长200cm的换气缝。在2000年1～2月（最冷月），当室外气温为-5～3℃时，该阳台温室气温可维持在11～18℃；当室外气温为-3～7℃时，该阳台温室气温可维持在12～21℃。无种植的阳台阳光间可节能约40%。

8.2.2 Balcony greenhouse effect experiment in winter

Example: In Xi'an City at a sun-facing balcony installed a sliding window with single glazing, inside the glass hanged a layer of agri-transparent-plastic film, up the window leaved a vent of 5cm height by 200cm length. In the coldest months January～February, 2000, when the outdoor air temperatures were -5～3℃, in the balcony air temperatures were kept at 11～18℃; outdoor air -3～7℃, in balcony air 12～21℃. No plant balcony sunspace can save energy about 40%.

2007年2月又一次实验（无塑料薄膜）。

实测数据如下：

On Feb/2007, an experiment again (without the plastic film).

The measured data were as follows：

时间 Time 测点 Site	2007/02/17/15时，多云 15h/17/Feb/2007Cloudy	2007/02/20/15时，多云转晴 15h/20/Feb/2007Cloudy→Fine	2007/02/23/15时，阴转晴 15h/23/Feb/2007Overcast→Fine
阳台 Balcony℃	25.0	27.0	28.0
主室 Room℃	23.0（供暖 Heated）	25.0（供暖 Heated）	25.0（供暖 Heated）
室外 Outdoor℃	20.0	21.0	22.0

从该数据可看出，房间不仅无失热，相反还从阳台（温室）得热。温室的确是一座节能的好空间。

From the data we can see that the room had no any heat loss, inversely, heat gain from the balcony greenhouse. Greenhouse does be a good energy saving space.

8.2.3 种植温室与日光间夏季比较

我们先用单层透明塑料薄膜围成0.5m×0.5m×1.35m的空间，阳光下室外气温35℃，温室（日光间）内温度达到49℃，内外差14℃。用一层塑料薄膜围起来的空间为什么里面比外面高出14℃呢？原因是透明塑料薄膜也与普通玻璃一样有"透短吸长"的热工特性，可透过85%以上的来自太阳的短波辐射，而其内部地面、空气、支撑骨架等被加热后，对外的长波辐射则被薄膜吸收，将其中约一半的吸收热再以长波辐射形式返回内部空间，加热空气。同时，由于该空间被封闭，杜绝了空气对外的对流散热，这就是该内部空间气温为什么比外面气温高的原因，也是通常称的温室效应。

8.2.3 Comparing planted greenhouse with sunspace in summer

We first used a layer of transparent plastic film to envelop a space of 0.5m×0.5m×

8 建筑绿化
8 GREENING OF BUILDINGS

1.35m to form a small sunspace. Under sunshine, outdoor air temperature was 35℃, inside the small sunspace air temperature was 49℃, the temperature difference was 14℃. Why a space enveloped by a thin plastic film can have it's temperature 14℃ higher than that of the outside air? The reason is that the transparent plastic film as same as the general glass has a character of "transmitting short wave radiation while absorbing long wave radiation". The film can transmit the solar radiation (short wave radiation) more than 85% into the sunspace, the ground, air and support are heated, they emit long wave radiation which is absorbed by the film, about half the radiation heat returns to the sunspace to heat the air again. At the same time the sunspace is closed so that can stop the outward convective heat loss. That's the reason why the temperature of the air inside the sunspace is higher than that of the air outside the sunspace. That's generally so called the greenhouse effect.

图 8-10(a) 是用同样体积的透明塑料薄膜围封一棵绿叶树作为有种植的实验小温室，测温仪清楚显示温度高达 57.3℃；当时室外气温与前述小日光间实验时相同，也是 35℃，此处内外温差高达 22.3℃，比前述小日光间实验时内外温差 14℃ 还高 8.3℃。

我们还进行了 6m 宽×12m 长×4m 高有种植的大温室夏季实验，当室外气温为 35℃ 时，温室顶部气温也达到了 57.3℃，见图 8-10(c)。

Fig. 8-10 (a) shows the same volume space enveloped by the transparent plastic film contained a tree of green leaves as a planted small experimental greenhouse, the temperature meter was clear showing the temperature being 57.3℃, while the outdoor air temperature was the same 35℃ as in the above small sunspace experiment. Here the temperature difference high reached 22.3℃ between inside and outside the space which was 8.3℃ higher than that temperature difference 14℃ in the above experiment of small sunspace.

We have done a summer experiment in a big planted greenhouse of 6m width×12m length×4m height, when the outdoor air was 35℃, the greenhouse top space reached 57.3℃ too, Fig. 8-10 (c).

从图 8-10(a) 还可看出，上部枝叶已枯黄(实验时间只有 4 个小时)，这又是什么原因呢？原因是这样的：绿色植物在阳光下进行光合作用，吸收空气中的二氧化碳与根系输送来的水，通过叶绿素变成有机物碳水化合物，同时放出氧气。植物也和其他生物一样有呼吸作用，吸氧，呼出二氧化碳。植物吸入的氧气(约为其光合作用产氧量的 1/20)用为将叶绿素制成的有机物中的一小部分进行氧化，生成二氧化碳(远小于光合作用吸收的 CO_2)和水，并放出热能。现在用有机物葡萄糖($C_6H_{12}O_6$)为例，植物吸入氧气，将葡萄糖 $C_6H_{12}O_6$ 氧化，分解为二氧化碳和水，并释放热能，方程式如下：

$C_6H_{12}O_6+6O_2 \longrightarrow 6CO_2+6H_2O+2822kJ$ 热能，这些释放出来的热能除供植物生理活动需要一部分外，剩余的热能将释放到空气中，这就是有种植的温室内气温比无种植的温室内气温更高的原因之一。我们再看图 8-10(a) 所示温室上部，塑料薄膜内表面已结满了凝结水。这些水分，一部分由根部种植层里的水蒸发而来，其余乃由叶面蒸腾而来。在蒸发、蒸腾过程中，它们都是高温水蒸气，一遇到比温室气温 57.3℃ 低得多的塑料膜内表面 38℃ 左右温度，必然产生凝结水。高温水蒸气变成低温凝结水必然放出大量热能(相变

8.2 建筑绿化的含义·效益·实例·实验
8.2 The Meaning · Benefits · Examples · Experiments of Greening of Buildings

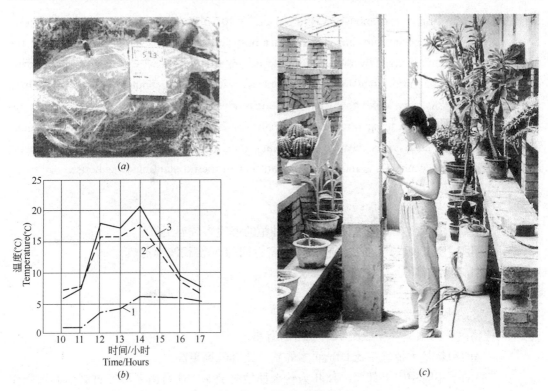

图 8-10 温室效应实验

Fig. 8-10 Greenhouse effect experiment

(a)有种植小温室夏季实验；(b)冬季有、无种植的温室效应比较；(c)研究生王进正在测大温室温度

(a) The experiment of a small planted greenhouse in summer; (b)Greenhouse effect comparison between planted greenhouse and no-planted one; (c)Postgraduate Wang Jin is measuring the temperatures of the big greenhouse

1—室外气温 outdoor air temp.；2—无种植温室气温 no-planted greenhouse air temp.；

3—有种植温室气温 planted greenhouse air temp.

(1992年1月28日，多云)(1992,Jun,28,cloudy)

潜热）。这就是有种植的温室内气温为什么更高的又一原因。

From Fig. 8-10 (a) we can also see in the greenhouse the tree's up part leaves already were scorched (the test time only 4 hours), why? The reason is that: the green plants under sunshine carry on their photosynthesis, absorb CO_2 from air and absorb water from root system, they through chlorophyl become organism, carbohydrate while release O_2. Plants also have breath as other living things, breathe in O_2 and out CO_2. Plants breathe in O_2 (about one twentieth of the outputting O_2 by photosynthesis) to oxidize a few parts of the organism made by chlorophyll to produce CO_2 (far less than the CO_2 absorbed by the plant in photosynthesis) and water while release heat. Now take the organism glucose ($C_6H_{12}O_6$) as an example: the plant breathes in O_2 to oxidize the glucose($C_6H_{12}O_6$), decomposes it into CO_2 and water and releases heat, the equation is as follows:

$$C_6H_{12}O_6 + 6O_2 \longrightarrow 6CO_2 + 6H_2O + 2822kJ \text{ heat energy.}$$

One part of the released heat energy satisfies the plant physiologic needs, the remains of the heat energy will be released into air, that's one of the reasons causing the air temperature in the planted greenhouse being higher than that of the air in the no planting greenhouse. Let's see Fig. 8-10 (a) again, it shows that in the greenhouse upper part, the plastic film inner surface has been full of condensed water: one part of the water from the planting layer by evaporation, the others from the leaves' transpiration. During evaporation and transpiration they are high temperature water vapour, contact the plastic film inner surface of about 38℃ very lower than that of the greenhouse air 57.3℃, certainly the conditions cause condensation. High temperature water vapour condensed into low temperature water, must release more heat energy (phase change latent heat), that's another reason why the air temperature in the planted greenhouse is more higher than that in the no planting greenhouse.

8.2.4 小结

从上述实验可看出，有四种原因造成有种植的温室内气温更高：
1) 透明膜"透短吸长"的热工特性所决定返回的50%长波辐射热；
2) 透明膜封闭后杜绝了向外对流失热；
该1)、2)两种热与日光间相同。
3) 蒸发的水蒸气变成凝结水放出的潜热；
4) 植物蒸腾的水蒸气变成凝结水放出的潜热。
3)、4)两种热是种植温室比日光间多的热，故其气温更高。

在实验的4小时中，上部树的枝叶为什么都枯萎了呢？这是由于温度过高引起蒸腾作用过快，根系水供不上造成的。但从57.3℃的高温可看出，有种植的温室内外热压差比无种植的温室内外热压差更大。说明，如果在有种植的温室上下设立适当的通风孔，就可能做到通风降温，更换新鲜空气，并保持植物最佳光合温度状态。大温室，由于根部供水足够，枝叶未枯黄。另外，由于大温室空气湿度增高引起空气中水蒸气压力加大也会阻止蒸发和蒸腾作用。

8.2.4 A brief summing-up

From the above experiments we see four reasons causing the air temperatures more higher in the planted greenhouse:

1) the returned 50% long wave radiation heat by the transparent film's character of "transmitting short wave radiation while absorbing long wave radiation";

2) the space enveloped by the film stops the outward convective heat loss;

The heat of 1) and 2) is as the same in sunspace.

3) the evaporated water vapour becomes condensed water while releases the latent heat;

4) the water vapour from the plant leaves transpiration becomes condensed water while releases the latent heat.

The heat of 3) and 4) is the more heat in planted greenhouse than that in sunspace, so the air temperature in planted greenhouse is more higher.

During the four hours of experiment why the upper branches and leaves of the tested tree were withered? That's due to the high temperature speeding transpiration while the water supply can't follow up from its root system. But we can see that from the high temperature of 57.3℃ the thermal pressure differential between the inside and outside of the planted greenhouse is higher than that of no planting greenhouse that's showing us, if at the top and bottom of the planted greenhouse set suitable vents, we can get ventilation cooling and fresh air exchange and keep an optimal photosynthesis temperatural situation. In the big planted greenhouse the water supply from root system was enough so the branches and leaves were not withered. On the other hand, because in the big greenhouse the air humidity increased, it caused increase of air steam pressure which can stop the evaporation and transpiration.

8.2.5 种植温室与日光间冬季比较

用 $1m^2$ 见方，2m 高的两个同体积透明塑料膜，用轻型木构架支撑，一个罩一棵常青树作为有种植的温室，另一个内部空着，作为无种植温室（阳光间）。图 8-10(b)为实验记录的该有种植温室与阳光间温度比较曲线图。从图中明显看出，在室外 1.5～6.5℃多云天气情况下，有种植的温室气温白天都高于无种植的温室气温。最高时有种植温室气温达到 21℃，无种植温室气温为 17.5℃。再次证实，有种植的温室效应优于无种植的温室效应。有种植的阳台温室可节能 60%左右（阳光日）。

8.2.5 Comparing planted greenhouse with sunspace in winter

With transparent plastic film we enveloped two spaces of the same volume of $1m^2$ area×2m height, supported by wooden light structure. In one of the two enveloped spaces an evergreen tree was inside it as a planted greenhouse, another emptied as a no planting greenhouse (sunspace). Fig. 8-10 (b) is the comparison of temperature curves between the two. We see clear, during outdoor 1.5～6.5℃, cloudy, the planted greenhouse air temperatures were all higher than that in the sunspace whole daytime. The maximum temperature in planted greenhouse was 21℃ while in the no planting greenhouse was 17.5℃. It's proved again the planted greenhouse effect is better than that of the no planting greenhouse. A planted balcony greenhouse may get a energy saving of about 60% (sunshine days).

8.3 建筑绿化构造设计

8.3.1 屋面绿化构造

(1) 不同地区的屋面绿化

图 8-11 所示为昆明市某住宅平屋顶绿化。

昆明夏季不热，冬季无严寒，春秋温暖。宜人气候长，一年约 300 天，有"春城"美称，年平均相对湿度为 73%，年降雨量超过 1000mm，所以昆明很适合花草树木生长，现已成为国内外商品花供应大站之一。图 8-11 中屋顶花草树木生长茂盛，其剖面基本构造：承重层（钢筋混凝土板）、水泥砂浆找平层、防水层、水泥砂浆保护层及 50cm 的土种植层。种植物浇灌用自来水。四周泛水构造与传统建筑构造相同。

昆明气候也有异常的时候，如前所述，1983 年 12 月 27～29 日大雪深 30cm；2000 年

8 GREENING OF BUILDINGS

1月29～30日大雪积深亦达30cm（受拉尼娜气候灾害影响）。2008年，中国遭遇雪暴灾，昆明也受到影响。

8.3 Construction Design of Greening of Buildings

8.3.1 Roof greening construction

(1) Roof greening in different regions

Fig. 8-11 shows a flat roof greening of a dwelling, Kunming City, Yunnan Province.

In Kunming, the weather is not hot in summer and not very cold in winter, spring and autumn are warm. The comfortable weather may last about 300 days yearly, she has a good name of "Spring City", annual mean relative humidity is 73%, annual rainfall over 1000mm, so in Kunming, the flowers, grasses and trees are suitable to grow. She has become one of the big stations supplying commercial flowers inside and outside China. In Fig. 8-11, on the roof the flowers, grasses and trees grow in flourish, the roof section basic construction: bearing member (RC slab), leveling coat of cement-sand mortar, waterproof course, cement-sand protective coat and the planting layer of 50cm soil. The plants are watered by tap water. The surrounding flashing is as the same as that of conventional building.

Kunming climate some times is abnormal, as said above, in 1983 December 27～29, in Kunming it snowed 30cm; in 2000 January 29～30, it also snowed 30cm (it was affected by the LaNina climate disaster). In 2008, China suffered snow storm's damage. Kunming also was affected.

图 8-12 是法国巴黎某旅馆平屋顶种植。由于巴黎冬季寒冷，故该屋顶在承重层上铺有绝热层，其上依次为保护层、防水层、保护层、50cm 土种植层。种植层下设有卵（砾）石滤水层10cm。

Fig. 8-12 shows a flat roof planting of a hotel in Paris, France. Paris winter is cold, on bearing course of the roof is a insulation, then in turn, protection, waterproofing, protection and 50cm planting layer, beneath this layer is a filter of 10cm pebbles (gravels).

图 8-11 云南昆明某住宅平屋顶绿化
Fig. 8-11 A flat roof greening of a dwelling in Kunming City, Yunnan Province

图 8-12 法国巴黎某旅馆平屋顶种植
Fig. 8-12 A hotel roof planting, Paris, France

8.3 建筑绿化构造设计
8.3 Construction Design of Greening of Buildings

彩图 8-13 是上海市卫星城之一嘉定某宾馆屋顶绿化例。这是一幢平顶与双坡顶的组合结构,在平顶部分设钢筋混凝土花架和种花槽,内填土 50cm 为种植层。由于嘉定气候温湿,相对湿度都在 77% 以上,年降雨超过 1100mm,故花草树木极易生长。彩图 8-13 显示蔷薇正盛开,已攀藤到坡顶,并向前延伸。

Color picture 8-13 shows a hotel roof greening, Jiading Town, one of the satellite towns of Shanghai City. This roof' structure is a combinative structure of a flat roof and a double-pitch roof. On the flat roof set a RC frame for flowers to climb and a plantable trough filled with 50cm soil as planting layer. Jiading climate is warm-wet, the relative humidity is general above 77%, annual rainfall over 1100mm, so the flowers, grasses and trees are very easy to grow. Color picture 8-13 is showing the rose being in full blood, climbing up the pitch roof and spreading forward.

(2) 自然绿化实验

彩图 8-14 是作者在西安进行的掩土太阳房自然绿化实验。什么是自然绿化呢?我们知道、也见过,野生植物、花草生命力很强,所以生长良好。这种现象启发我们进行了这次自然绿化实验。该房为 24 砖外墙,向阳面有窗。冬季通过窗户太阳直接得热,其他东、西、北三面覆土。屋顶为现浇钢筋混凝土平屋顶,2cm 水泥砂浆找平层涂防水涂料,其上为 30cm 种植层(50% 黄土拌 50% 锯末)。让飞来的植物种子自然生长。三年后,自然绿化即成彩图 8-14 状态。2000 年屋顶有的树高已 4m 多。西安年降雨量约 640mm,年平均相对湿度 71%,冬季除一些常青植物外,大多数均叶黄枯落,但第二年又自返绿再生,效果之好实出意料。16 余年一直无任何渗漏现象。

(2) A natural greening experiment

Color picture 8-14 shows a natural greening experiment on an earth sheltered solar building in Xi'an region by the authors. What means natural greening? We have known and seen the wildings growing well because of their strong life force. These phenomena enlighten us to do the natural greening experiment. The exterior wall of the building is 24cm clay brick wall with sun-facing windows. In winter can get solar direct gain through the windows, the other three faces with earth covered. The roof is a cast-on-site RC flat roof. On the roof in turn is a 2cm cement-sand leveling coat, a waterproof paint, and a 30cm plantable layer (50% loess mixed with 50% sawdust). The flown plant seeds naturally grow there. Three years later, the natural greening is as shown in color picture 8-14. In 2000, on the roof a few of the trees had grown up more than 4m tall. In Xi'an the annual rainfall is about 640mm, mean relative humidity 71%. In winters except the several evergreen plants, most plants are withered, but next year they all can renew themselves to return in green, the effect surprised us. More than 16 years, the roof is still having no any leakage.

全球的绿化,自然绿化才是第一主角!它能因地制宜顺当地天地之道优化出多元品种植物,形成当地生态可持续发展的植物群繁衍下去。

In the global greening the natural greening does be the No. 1 role! It can according to the local condition of climate and geology optimize multifarious kinds of plants forming the local sustainable developing plant groups to multiply forwardly.

圆明园遗址 150 多年的自然绿化已经优化出一个多元品种的乡土植被群,且饱含帝国

8 建筑绿化
8 GREENING OF BUILDINGS

强盗掠夺的痕迹,激发后人发愤图强的潜能。相反,搞什么恢复圆明园的人工再整形,实是逆天地之道的负功行为!

Yuan-ming-yuan remains' natural greening have been more than 150 years, have optimized a vernacular vegetation group of multifarious kinds, fully containing the robbed marks of empire robbers and the potential encouraging the offspring to encourage themselves being powerful. By contraries, to do what renewing Yuan-ming-yuan's artificial face-lifting that truly a negative doings.

(3) 平屋顶温室轻质种植

图 8-15 为一平屋顶温室剖面例。传统屋顶如前所述只能起单纯的围护作用。屋面种植后既利用了太阳能又利用了雨水。从前,屋面种植多用土种植层,一般厚 50cm,浇水后每平方米屋面增重就达 700 多千克,太重。有的曾用 25cm 深蓄水养殖屋面,夏季降温效果很好,但寒冷地区不适用,除非有良好的保温避免结冰。而且,25cm 深的水层,每平方米 250kg,仍然较重。前述的西安自然绿化实验,采用了黄土加锯末的种植层 30cm 厚,每平方米约 240kg,植物虽然生长甚佳,但增重仍不少。之后我们又实验了蛭石、锯末轻质种植层,厚仅 15cm,所种多种花卉如牵牛花(喇叭花)、茉莉、月季和夜来香都生长良好。15cm 厚蛭石或锯末浇水后每平方米增重仅 30～35kg,很轻。据此证明,屋面轻质种植比蓄土种植、蓄水养殖好;而屋面温室种植比开敞种植好,可延长绿化期,发挥生态正效应时间长。屋面温室对冬季保暖、节能大有好处,夏季将通风口开启,可最大限度发挥蒸发与蒸腾制冷的效应。轻质层比同厚度的土层或蓄水层所起的冬、夏绝热作用都大,更有利于节能。

(3) Light mass planting greenhouse on flat roof

Fig. 8-15 shows a section example of flat roof greenhouse. As above, the conventional roof can only do the single function of enveloping. If the roof is planted not only the solar energy can be used while the rainfall too. Before now, roof planting usually used a planting layer of 50cm soil, which watered will increase the roof load of more than $700kg/m^2$, it's too heavy. Some one used a 25cm deep breeding roof pool, summer cooling effect was very good, but in cold region it's not available if without a good insulation to avoid freeze. However, the 25cm water increases load of $250kg/m^2$ that's still too heavy. As above, the natural planting experiment used planting layer of loess mixed with sawdust, 30cm thickness, about $240kg/m^2$, although plants grew well, the increased load was still not light. Following these, we tested with vermiculite and sawdust as planting layer, 15cm thickness, in which the flowers such as morning glory, jasmine, Chinese rose and cordate telosma all grew well. 15cm vermiculite or sawdust watered increases load only $30\sim35kg/m^2$, very light. The above has proved that roof planting with light mass is better than with soil or breeding water pool; roof greenhouse planting is better than open planting. Greenhouse planting has longer green term and longer term of positive eco-effect. Roof greenhouse in winter greatly benefits insulation and energy saving. In summer, open the vents to maximize the effects of evaporation cooling and transpiration cooling. In winter or summer the thermal resistance effect of light material is better than that of the same thickness soil or water, and more benefiting energy

8.3 建筑绿化构造设计
8.3 Construction Design of Greening of Buildings

saving.

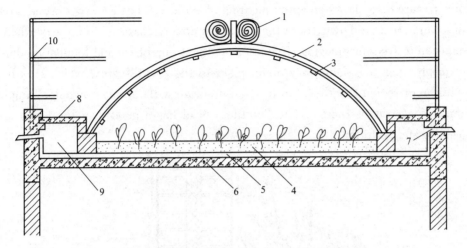

图 8-15 平屋顶温室剖面示例（设计：王进；指导：夏云）
Fig. 8-15 An example of flat roof greenhouse section (Designer：Wang Jin；Director：Xia Yun)
1—可动绝热帘 movable insulation curtain；2—透明塑料膜 transparent plastic film；
3—支架 support structure；4—蛭石轻质种植层 15cm vermiculite light mass planting layer,15cm；
5—防水层 waterproof coat；6—承重层 bearing slab；7—泄水管 overflow pipe；
8—3cm 厚预制钢筋细石混凝土走道板 walkway,3cm thick RC slab；
9—蓄水槽 water trough；10—防护栏杆 protective railing

8.3.2 散水绿化构造

(1) 散水种植剖面及实例

图 8-16 为散水绿化设计剖面例。彩图 8-17 为西安建筑科技大学散水绿化实验实景。由图 8-16 剖面可看出，散水种植很简单，在传统散水面上加填 200～300mm 土层作为种植层即可。200～300mm 厚种植层已足够种植一般花草及常青灌木。该种植层还可作为附加绝热层，使冬失热和夏进热减少，消除冷（热）桥作用。

彩图 8-18 为上海嘉定某宾馆散水种植常青灌木与季青乔木的实景。该地土质无湿陷或湿胀性，又无冻融循环性，故地基及基础无防水要求，植物可直接种在回填土上，生长良好。该房采用了坡顶平瓦屋面和自由落水，没有水落管，构造简、排水通畅。平瓦前沿滴水构造好，所以该檐口一直干净，无污染。

8.3.2 Apron greening construction

(1) Apron planting section and practical example

Fig. 8-16 shows an example of apron greening section design. Color picture 8-17 is a practical situation of a apron planting experiment in Xi'an University of Architecture and Technology. From Fig. 8-16 section we can see that apron planting is very simple, on conventional apron fill up a planting layer of 200～300mm soil that's okay, 200～300mm planting layer is enough to plant general flowers and evergreen bush. This planting layer is also an attached insulation to decrease winter heat loss and summer heat gain and eliminate

8 GREENING OF BUILDINGS

cold/heat (thermal) bridge effect.

Color picture 8-18 shows an apron planting of evergreen bushes and seasonal green arbores in a hotel, Jiading Town, there the earth has no any characters of wet-settlement or wet-expansion or freezing-thawing circulation, so the ground base and foundation don't demand waterproofing, the plants may direct plant in the backfill, grow well. This building uses a pitch roof roofed flat tiles, and free-fall drainage without any downpipe, simple construction, drainage unobstructed. The flat tile's head has a good drip, the eaves are still kept clear, no any pollution.

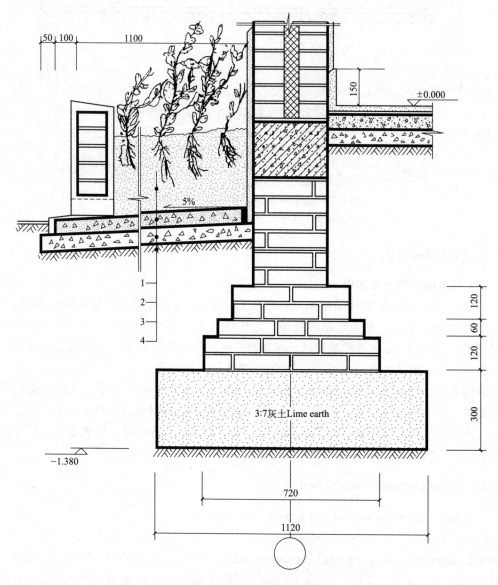

图 8-16 散水绿化剖面设计示例

Fig. 8-16 Apron greening section design example

1—300 厚土种植层 300t soil planting layer；2—60 厚混凝土 60t concrete；

3—80 厚碎石垫层 80t broken stone base；4—素土夯实 original soil ramed，(t=thickness)

8.3 建筑绿化构造设计
8.3 Construction Design of Greening of Buildings

(2) 檐口自由落水有利于散水种植

檐口自由落水会不会被风吹到墙面而污染墙面呢？通过对上海、南宁、成都、西安等地多次观察，从低层到多层，结果发现，檐口自由落水不但不被风吹向墙面，反而偏离墙面。通过实验进一步得到证明。图8-19(a)、(b)，图8-20为实验记录。图8-19(a)为用线绳替代水的风吹实验；图(b)为用水的实验。从两图均可明显看出，当风吹向墙面时，檐口自由落水(图(a)中檐口的线绳和图(b)中檐口的深色水线)反而偏离墙面；而天空降雨(图(a)中飘向墙面的线绳和图(b)中飘向阳台的白色水线)则直向墙面、阳台吹进。檐口自由落水的水滴直径达到6～8mm重力才能克服附着力拉檐口水下降。风速越大，则偏离墙面越远。图8-20示出了三图实验例。在背风面和其他面也同样存在檐口自由落水偏离墙面现象。

(2) Eaves free falling drops benefiting apron planting

Could eaves free falling drops be blown onto wall and pollute the wall? Through many times of observation in Shanghai, Nanning, Chengdu and Xi'an cities, from low-story to multi-story buildings we discovered when wind blew to the wall the eaves free drops were not blown onto the wall inversely deviated from the wall. The experiments have further proved this phenomenon. Fig. 8-19 (a)(b), Fig. 8-20 are the experimental records. Fig. 8-19 (a) shows the experiment of using thread instead of water in wind blowing; (b) shows the experiment using water. From the two figures we can clear see, when wind blows to the wall the eaves free fall-

图8-19(a) 檐口自由落水风吹偏离墙面，天空雨击向墙面。实验(1)(摄影者：罗升建)

Fig. 8-19 (a) When wind blows to the wall eaves free falling drops are deviated from the wall while sky raindrops are struck against the wall. Experiment(1)

(Photographer：Luo Shengjian)

(以线绳代水) (cotton rope instead of water)

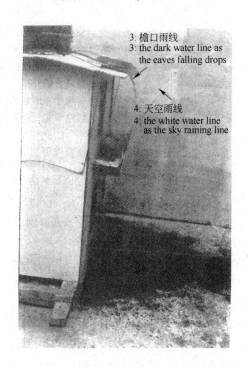

图8-19(b) 檐口自由落水风吹偏离墙面而天空雨却击向墙面。实验(2)(摄影者：罗升建)

Fig. 8-19 (b) When wind blows to the wall, eaves free falling drops are deviated from the wall while sky raindrops are struck against the wall. Experiment (2)

(Photographer：Luo Shengjian)

ing drops (in Fig. 8-19 (a) the eaves thread and (b) the eaves dark color water line) inversely are deviated from the wall while the sky raindrops (in (a) the thread blowing forward to the wall, in (b) the white water line blowing onto the balcony) are straight flying onto the wall and balcony. When the eaves drop's diameter reached 6~8mm its gravity can overcome the adhesive force to make the drop fall down. The more faster wind speed the more deviation caused. Fig. 8-20 shows three experimental examples. In the lee-side and other sides the eaves drops are also deviated from the wall.

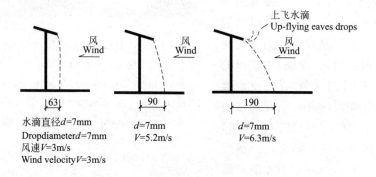

图 8-20 檐口自由落水风吹时偏离墙面。实验(3)

Fig. 8-20 When wind blows to the wall, eaves free falling drops are deviated from the wall. Experiment(3)

为什么檐口自由落水不管迎风面、背风面和其他面不湿墙反而偏离墙面呢？原来这是由一条普遍而又简单的流体动力学规律所决定的。人人都可自己动手做一个极简单的实验：

用左右手中、食指夹住信笺短边左右角，让其自然下垂，然后用嘴贴近其上沿向前吹气，信笺就会向前、向上飞起，如图8-21所示。原因是当信笺自然下垂，无风时，其两侧空气压力相等，故可保持自然下垂不动。当在左(上)侧吹气时，纸左(上)侧消耗的动能增大，同时静压能相应减小，而右(下)侧静压没有减少，因此右(下)侧空气静压大于左(上)侧空气静压，就将纸抬起来了。

做了上述实验，就可按图8-22来分析檐口水的自由下落规律了。

Why the eaves free falling drops in all cases of facing-wind, lee-side and other sides are not blown onto wall to wet the wall, inversely deviated from the wall? We discover that's controlled by a general and simple flow dynamics' law. Every one can do a very simple experiment by oneself:

Let your left and right hands' middle finger and index finger grip the two angles of the short side of a letter paper, let it naturally hang down, and your mouth close to the up-side of the paper, forward blow the air, the paper will fly upward and forward as Fig. 8-21 showing. The reason is that when the paper is naturally hanging down, no wind on both the sides of the paper, their air pressures are equal so it can keep the paper naturally hanging down and still. When blow the air on left (up) side the left (up) side air kinetic energy consumption is increased and its static energy (pressure) correspondingly decreased while

8.3 建筑绿化构造设计
8.3 Construction Design of Greening of Buildings

on the right (down) side, the static pressure is not decreased, which is bigger than that of the left (up) side, so the paper is lifted up.

Completed the above experiments, we may analyse the law of eaves drops free falling down with Fig. 8-22.

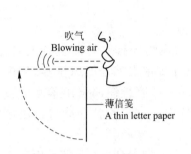

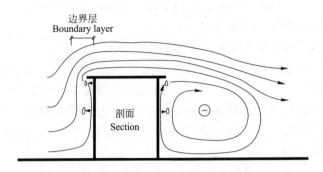

图 8-21 吹纸实验

Fig. 8-21 An experiment of blowing paper

图 8-22 檐口自由落水偏离墙面的风动力学分析

Fig. 8-22 The analyzing of eaves drops deviating from the wall by air dynamics

先看迎风面，当风吹向墙面，遇墙必然向上绕过檐口以及向两旁山墙绕过，风由向墙到转变方向顺墙流动。该顺墙流动的空气层我们称为边界层。边界层内已没有推水滴向墙的力量了。相反，水滴外侧空气顺墙流速快，静压就小，其内侧由于墙面的湍流边界效应和摩擦阻力，空气顺墙流速就慢，静压就大，因此将水滴推离墙面。流体力学早已告知我们："流体中的物体必向流速快的一侧移动"。海洋中相近两船行驶中发生相撞事故，原因也就在此。再看图 8-22 檐口水滴在未下落或刚下落时，还会遇到上翻风力的外推作用，也会使其偏离墙面。这就是迎风面檐口自由落水偏离墙面的原因。背风面檐口下落水滴也必然偏离墙面，只是偏离距离可能不同而已。同理，两端山墙侧风面，结果也是一样。檐口自由落水较有利于散水种植，可使屋面水均匀地浇灌散水所种植物。而且自由落水有更多的优点：排水畅通、构造简、配件少、施工易、造价低。

First observe the wind-facing side, when the wind blows to the wall, as reaching the wall it certainly will upward round the eaves and the both side gables, the wind will change its direction from toward the wall to along the wall. The air flow moving along the wall we call it boundary layer, in the boundary layer there is no any force to push the eaves drops onto the wall. Inversely outside the drops the air flow is faster than that of inside the drops. Inside the drops the air flow is slower due to the turbulence boundary effect and friction drag, so the static pressure is bigger, it pushes the drops outward from the wall. The fluid (flow) dynamics has already told us the law: "A body in fluid certainly moves to the faster flow side", in ocean two near seagoing ships in moving probably impact with each other that's caused by the law. See Fig. 8-22 again, when the eaves drop before falling down or just falling down, the upward wind will outward push the drop to deviate from the wall, that's the reason in wind-facing side why the eaves free

8 建筑绿化
8 GREENING OF BUILDINGS

falling drops are deviated from the wall. In lee side the eaves free falling drops are deviated from the wall too, the deviation distances may be different. In the same reason in both sides of the sidewalls all the results are the same. Eaves free falling drops benefit apron plants, it can make the roof water to equally water the apron plants. Besides, eaves free falling drops have more advantages: clear drainage, simple construction, less fittings, easy to do and low cost.

有水落管排水、自由落水哪个好？

传统观念里有水落管排水称有组织排水，自由落水为无组织排水，此观点欠妥。因为，屋面水向一面、二面或四面排都必须进行设计（设计排水也就是组织排水）。中国各民族传统建筑和广大农村建筑绝大多数都是有设计的，即有组织的自由落水。我们认为，屋面排水都是有设计即有组织的，只是如何排到地面有水落管和檐口自由落水之分。

有水落管排水的缺点：易堵塞，配件多，构造繁，施工难，造价高。在湿胀或湿陷性地区，有水落管排水，很多并未引入下水道，而是排到散水再渗入地下。该渗入地下的水是不均匀的，易引起房屋不均匀下沉。

用塑料绳或铝链引水代替水落管排水，既无冻裂管子的危险，施工与维修也较方便，并可作攀藤植物的支撑。图 8-23 为铝链引水例（法国）。

Downpipe drainage, free fall drainage, which is better?

In conventional concept, roof drainage with downpipe is called organized drainage, free fall called unorganized drainage, this concept is not reasonable because the roof water drained off by one direction or two or four we should carefully design (to design drainage i. e. to organize drainage). In China most roof drainages of every national conventional buildings and the vast vernacular buildings are designed i. e. organized to free falling. As our opinion all roof drainages are designed drainage i. e. organized drainage, the only difference is how to drain the roof water to ground: with downpipe or free fall.

Downpipe drainage's shortcomings are: easy to be obstructed; more fittings, complicated construction, hard to do and high cost. In wet-expansion or wet-settlement region, most buildings with downpipe drain the water not into the sewer only onto apron then let the water permeate ground. The water permeated ground is not a uniform distribution, it easy causes the building's un-

图 8-23 用铝链代水落管引水入下水道（法国）

Fig. 8-23 Aluminium chain instead of downpipe to lead roof water into the sewer (France)

8.3 建筑绿化构造设计
8.3 Construction Design of Greening of Buildings

even settlement.

Using plastic rope or aluminium chain instead of downpipe to lead roof water not only can avoid the downpipe to be broken by freezing, the construction and maintenance are also convenient and the rope or chain also can be the supports of climbers. Fig. 8-23 shows an example of using aluminium chain to lead the roof water into the sewer (France).

8.3.3 窗台、阳台绿化

(1) 阳台种植实验与实例

彩图8-24(a)、(b)为医学教授张宜仁与作者的阳台种植实验。采用"山"形双槽形窗台作为种植槽，中间用推拉窗分隔内、外槽，外槽种季青攀藤植物，夏季遮阳，冬季阳台内可得阳光。内槽植物，由于温室效应可延长冬季绿化期，并有利于来年再生。

彩图8-25为深圳市某滨海宾馆窗台花盘种植景观。该宾馆位于蓝天碧海的绝佳境地，可以推想，待到盘花盛开时，衬在白色建筑上，一座立体花园带着美景巍然而立。

莫斯科冬季又长又冷，居民仍爱在阳台上设槽种花。图8-26为莫斯科某住宅在阳台上采用木槽种花的情况。

图8-27为克罗地亚萨格勒布市挑阳台绿化例。

8.3.3 Windowsill and balcony greening

(1) Balcony planting experiments and examples

Color pictures 8-24 (a),(b) are the medical prof. Zhang Yiren and the authors' balcony planting experiment. We use a double trough "山" shaped windowsill as the planting trough, with a sliding window to partition the inside trough and outside one, in outside trough plant seasonal green climbers which can get summer sunshade and winter sunshine. In inside trough due to greenhouse effect the plants can last winter green term and benefit the next year renewal.

Color picture 8-25 shows the view of windowsill planting at a seaside hotel, Shenzhen City, South China. This hotel is located in a very beautiful environment with clear blue sky and dark blue sea, we may imagine as the windowsill flowers being full blooming set off by the white building, a stereoscopic garden with beautiful sight will majestically stand there.

Moscow winter is very cold and long, the inhabitants also love balcony planting with trough. Fig. 8-26 shows in Moscow, a dwelling balcony planting with a wooden trough.

Fig. 8-27 shows an overhang balcony greening, Zagreb City, Croatia.

彩图8-28为一西方古典建筑外窗台绿化例。

从以上诸例可看出，窗台、阳台绿化早已有之，人人喜爱。

Color picture 8-28, an outer windowsill flowers of a western classical building.

From the above examples, we may see that the greening of windowsill and balcony has been already existed, every one likes it.

8 建筑绿化
8 GREENING OF BUILDINGS

图 8-26 莫斯科一居民住宅楼阳台木槽种花

Fig. 8-26 A dwelling balcony wooden trough planting, Moscow City

（2）窗台、阳台绿化剖面构造

图 8-29(a)、(b)为阳台、窗台绿化剖面设计例。(a)图为阳台温室种植。双槽种花槽设在离阳台地面90cm高处,其下空间可沿阳台地面周边种植其他花卉。根据实验,内槽花卉即使在冷区,由于良好的热湿温室效应,很多花卉可保持常青状态。外槽则可种季青爬藤植物,如牵牛花(彩图 8-24(b)),热季可增强遮阳;冷季,室内可得阳光。(b)图为窗户温室剖面设计例(详见图下说明)。

（2）Construction sections of greening of windowsill and balcony

Fig. 8-29 (a), (b) are the section design examples of greening of windowsill and balcony. (a) is a balcony greenhouse planting, the double trough planting trough set at the 90cm level from the balcony floor, below the level may plant other flowers on the balcony floor along the sides. According to the experiment, although in cold region, many flowers in inner trough can keep evergreen because of the good greenhouse effect of heat and wet, in the outer trough may plant seasonal green climbers such as morning glory (Color picture 8-24(b)), the climbers can increase sunshade in summer and gain sunshine inside room in winter, (b) is the section design example of window greenhouse (detail below the picture).

图 8-27 克罗地亚萨格勒布市挑阳台绿化例

Fig. 8-27 An overhang balcony greening, Zagreb City, Croatia

8.3 建筑绿化构造设计
8.3 Construction Design of Greening of Buildings

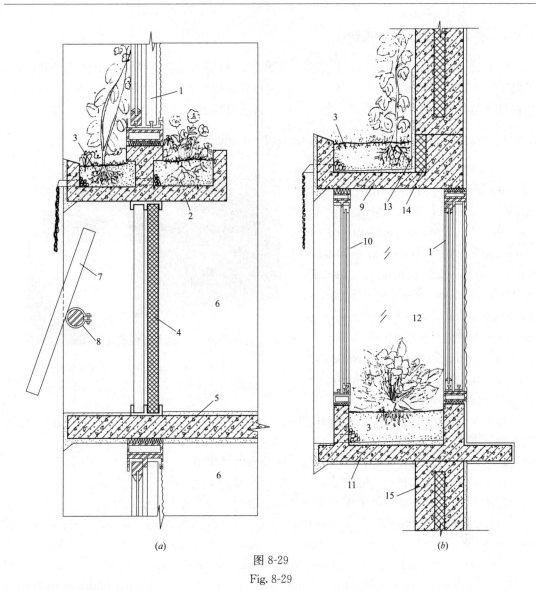

图 8-29
Fig. 8-29

(a)阳台绿化剖面设计 Balcony greening section design；(b)窗台绿化剖面设计 Windowsill greening section design
(设计者：夏葵)(Designer：Xia Kui)

1—带纱窗推拉玻璃窗 Sliding window with screen window；2—阳台双槽形种植槽，每槽净空 200mm×300mm 宽 Double trough planting trough, each trough net space 200mm×300mm width；3—蛭石或珍珠岩种植层 150mm Vermiculite or pearlite (perlite) planting layer 150mm；4—种植槽下推拉绝热板，高 900mm，厚 100mm，无毒泡沫塑料板，外包 0.2mm 厚铝板保护 Sliding insulation beneath planting trough, 900mm height, 100mm thickness, nonpoisonous formed plastics enveloped by a 0.2mm thickness aluminium sheet as a protection；5—钢筋混凝土阳台挑板 Overhang RC balcony slab；6—阳台垂直分隔板 Vertical balcony partition；7—太阳能电池板或太阳能热水器 Solar cell panel or solar hot water collector；8—可调支承轴，两端与垂直分隔板作结构联结 Adjustable support bar of two ends joined up with the vertical partitions；9—窗过梁或圈梁挑槽，净空 200mm×300mm 宽 Overhang trough from lintel or girth, net space 200mm×300mm width；10—推拉玻璃窗 Sliding window；11—窗台种植槽净空 200mm×400mm 宽 Windowsill planting trough of net space 200mm×400mm width；12—窗户两端推拉窗 Sliding windows at the both sides of the window；13—防水砂浆抹面 20mm×750mm 高 Waterproof mortar finish 20mm×750mm height；14—附加绝热层，减轻冷(热)桥作用 Attached insulation to reduce thermal bridge effect；15—高绝热复合墙板 High insulated composite wall

8 建筑绿化
8 GREENING OF BUILDINGS

8.3.4 墙面、柱面、挑檐面绿化

图 8-30 为某市市政府大门绿化实景。攀藤植物利用其气根的吸附作用和分泌的粘液,可以攀附在砖墙、光滑大理石板材、瓷砖、挑檐底表面等处。我们曾见到散水种植的爬藤植物可攀藤到六层楼檐口那么高,冬季落叶,来春自绿再生。

8.3.4 Surface greenings of wall, column and overhang eaves

Fig. 8-30 shows the greening of a gate of a city government. The climbers using their aerial root adsorption and the adhesion of the mucus secreted can climb and adhere to the surfaces of brick wall, smooth marble panel, ceramic tile and the bottom of cornice (overhang eaves; projecting eaves), etc. We have seen an apron climber climbing up the cornice of a six-story building, in winter leaves will be withered and fallen down, next spring will get self-green and renewal.

图 8-30 某市市政府大门绿化实景

Fig. 8-30 A city government gate greening

彩图 8-31 为国外一别墅,墙面攀藤红玫瑰,庭院有白、黄二花,红、白、黄三色交相辉映,构成一美丽的庭院花园。

彩图 8-32 为北京大学某教学楼墙面、屋面攀藤绿化情景。

彩图 8-33 为匈牙利布达佩斯某使馆凹阳台及墙面绿化。

彩图 8-34 为西安市园艺工作者为庆贺国庆节制作的花卉九龙壁。

Color picture 8-31 shows an outland villa, red rose climbing up the wall, white and yellow flowers growing in the court yard, the colours of red, white and yellow adding radiance and beauty to each other, all these form a very nice court garden.

Color picture 8-32, climber greening on wall and roof of a teaching building in Beijing University.

Color picture 8-33, greening of the recessed veranda and wall of a embassy (diplomatic mission) Budapest, Hungary.

Color picture 8-34, the Nine-dragon-wall made of flowers by Xi'an City gardening workers for National Day.

8.3.5 围墙绿化

彩图 8-35、彩图 8-36 分别为北京、深圳花卉围墙例。这与传统的砖砌围墙上装铁丝网两者不可同日而语。

8.3.5 Greening of enclosure

Color pictures 8-35 and 8-36 show the enclosures made of flowers in Beijing and Shenz-

8.3 建筑绿化构造设计
8.3 Construction Design of Greening of Buildings

en separately, comparing with the traditional enclosure built with brick and set iron netting on it, both the two cannot be mentioned with the same breath.

8.3.6 室内绿化

彩图 8-37 为广州白天鹅宾馆室内绿化。它的突出优点就是一个"真"字，真花、真树、真的小瀑布流入小清潭，很有中国园林风韵。

8.3.6 Indoor greening

Color picture 8-37 shows the indoor greening of White Swan Hotel, Guangzhou City, its great advantage may be shown with one word "true": true flowers, true trees, true small falls falling down the small limpid pool. It shows a high level graceful bearing of Chinese garden.

8.3.7 遮阳绿化

彩图 8-38 为上海嘉定西向商住楼西侧种的防西晒杉树，树已高过五层楼顶。彩图 8-39 为西安一西向住宅楼西侧种的高大白杨已达六层楼高。该杨树为季青乔木，夏叶茂密，增强遮阳，冬叶枯落增强室内日照。

彩图 8-40 是清华大学一阶梯教学楼东侧防东晒的季青绿化。作者夏云在该校学习时曾是该楼施工工地主任。

～·～·～·～·～·～·～·～

建筑及其环境绿化应注意：
季青与长青相结合；
高乔与低灌相结合；
美化与除害相结合。

8.3.7 Greening of sunshade

Color picture 8-38 shows a shopping-dwelling building's west side plants of firs for west sunshade. The firs have grown up over fifth-story of the building, Jiading Town, Shanghai City. Color picture 8-39 shows the west side tall poplars for west sunshade of a west-facing dwelling, Xi'an City, Shaanxi Province, the height of the poplars has been reaching sixth-story of the building. Poplars are seasonal green arbores, in summer thick leaves will increase of sunshade, in winter the leaves withered and fallen will increase the indoor sunshine.

Color picture 8-40, the seasonal green trees for east sunshade at the east side of a lecture theatre of Qinghua University. The author Xia Yun was the construction head during the building under building when he studied in the university.

～·～·～·～·～·～·～·～

Greening of buildings & their environments should consider:
Seasonal greens combining evergreens;
Tall greens combining short bushes;

8 GREENING OF BUILDINGS

Beautification combining eliminating harms.

8.3.8 不是一切建筑都绿化

彩图 8-41、彩图 8-42 分别为巴黎圣母院和非洲某教堂视景。一望便知,像这样高级的艺术建筑,如果谁还想用建筑绿化将它们掩饰起来,那不是太无情、太不懂建筑了吗!

北京的天安门、西安的钟楼以及其他历史性建筑、艺术高的建筑,都应保持其原有的面貌。我们主张建筑绿化,但并不是一切建筑都要绿化。

8.3.8 Not all buildings must be greened

Color pictures 8-41 and 8-42 show the sights of Notre Dame de Paris and an African church separately. By a looking at once we know these buildings with so high level art, if somebody attempts to cover them with greening the man might be having no any mercy and no any knowledge of understanding architecture!

Beijing Tiananmen Building, Xi'an Bell Tower and other historic buildings and high art buildings should be kept in their original style. We advocate greening of buildings but not make all buildings green.

9 优化组合
9 OPTIMAL COMPOSITIONS

9.1 能量自足房

20世纪90年代，欧洲的瑞士、德国建成了几幢不用常规商业能源的建筑，特称"能量自足房"。

图9-1为瑞士该房一例。

设计要点：

1) 主动系统与被动系统相结合，利用太阳能。屋顶装有30m² 太阳能电池板和25m² 太阳能热水器；

2) 外墙：30cm厚砖砌体与防水木丝板夹旧报纸制的高绝热层组成的复合墙；

3) 节能窗：采用三玻木窗，玻璃间充氪气，玻璃向室内的一面有透明热镜涂层，可减少2/3失热；

4) 所有钢筋混凝土梁板不插入外墙，消除冷桥效应；

5) 高效余热回收装置，70%空气余热可回收再利用。

9.1 Energy Self-sufficient House

In 1990s, in Switzerland and Germany, several houses were built without using normal commercial energy. They were especially named as "Energy self-sufficient house".

Fig. 9-1 is an example of this house in Switzerland.

Main features in design are:

1) Use solar energy by combining active system with passive system. There are 30m² of solar cell panel and 25m² of solar hot water collector on the roof;

图9-1 瑞士能量自足房
Fig. 9-1 An energy self-sufficient house, Switzerland

2) External wall: A composite wall consists of a layer of high performance insulation made of old newspaper, set between a 30cm thick brick wall and a waterproof wood fibre board;

3) Energy saving windows: Use triple-glazing timber windows, cavities between the glazing filled with krypton, on the surface of glazing facing room has heat mirror coating which can reduce 2/3 heat loss;

4) No RC beam or slab has been laid into exterior wall to eliminate cold bridge effect;
5) High efficient waste heat recycle system can recover and reused 70% of wasted heat.

9.2 透明保热墙

德国人在零能建筑中发明并利用了透明保热墙(TIW)。图 9-2 为工作原理，图 9-3 为其构造。

9.2 Transparent Insulated Wall (TIW)

A transparent insulated wall (TIW) was invented and used in a zero-energy house in German. Fig. 9-2 shows its operating principle. Fig. 9-3 shows its construction detail.

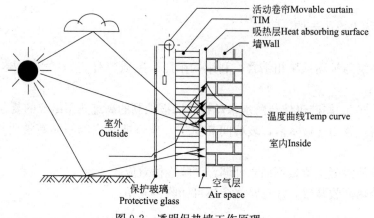

图 9-2 透明保热墙工作原理
Fig. 9-2 Transparent Insulated Wall (TIW) operating principle

TIW 与 Trombe 墙显著不同点是：TIW 中在黑色吸热表面与保护玻璃之间插入了一层透明绝热层(TIM)，厚 10cm。吸热面层不仅有透过的直接辐射(含天空云层反射)，还有 TIM 的反射。Trombe 墙只有一层空气层，TIW 有两层空气层，其热阻比 Trombe 墙大得多。黑色吸热面在 TIM 后面，使建筑立面艺术也得到相应改善。保护玻璃与 TIM 之间设有卷帘遮阳，可调节到达墙面的太阳辐射，以免大晴天室内过热，同时也增加夜间绝热。TIW 也有时间延迟性，与窗户直接得热相配合，可得到日夜较稳定的太阳能供暖。
图 9-4 为德国建成的 TIW 太阳房例。

The obvious difference between TIW and Trombe wall is that: in TIW, a layer of 10cm thick transparent insulated material (TIM) is installed between the heat absorbing surface and the protective glazing. The absorbing surface not only receives the directly transmitted solar radiation (incl the reflection of cloud) but also the TIM's reflection. Trombe wall has only one air gap but TIW has two air gaps. Therefore its thermal resistance is much bigger than that of Trombe wall. Because the black absorption surface is behind the TIM, the building façade's appearance can be improved accordingly. Between the protective glazing and the TIM has a roll of shading curtain which can adjust the amount of solar radiation reaching at the wall so as to avoid overheating in very fine days and increase night insulation. TIW also has time delay, which combined the direct gain through win-

9.3 透明保热层多种用途
9.3 Multi-use of TIM

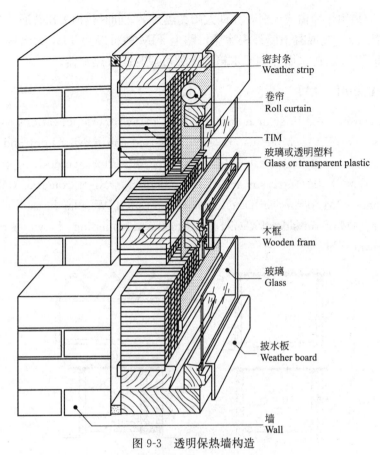

图 9-3 透明保热墙构造
Fig. 9-3 TIW construction detail

dows, may provide rather stable heating during daytime and nighttime.

Fig. 9-4 shows a TIW solar house example built in Germany.

图 9-4 德国建成的一幢透明保热墙太阳房
Fig. 9-4 A completed TIW solar house, Germany

9.3 透明保热层多种用途

TIM 除可用于太阳房外，还有多种其他用途（图 9-5）：

9 OPTIMAL COMPOSITIONS

图 9-5(a)、(b)用作墙窗或天窗，允许太阳光热入室，同时热损失比普通窗会大大减少。

图(c)，将旧砖墙或混凝土墙外表涂黑，贴上 TIM 就可成为 TIW 太阳房。

图(d)，将 TIM 贴在水墙外侧或水箱周围就可成为高效热水器。

9.3 Multi-use of TIM

Besides it can be used in solar house, TIM has many other applications (see Fig. 9-5).

(a)、(b) show TIM in windows and skylights that can let day light and heat coming into rooms while the heat loss is much less than that of normal window.

(c) shows that if the outer surface of the old brick wall or concrete wall be painted black, then adhere TIM to the wall, the house becomes a TIW solar house.

(d) shows TIM on outside of a water wall or the surrounding of a water tank, they can be an efficient water heater.

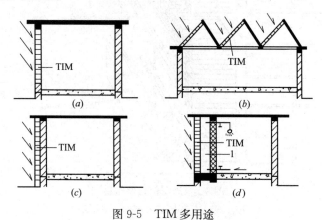

图 9-5 TIM 多用途

Fig. 9-5 Multiple-use of TIM

1—水 Water

9.4 自控光热屋顶

该屋顶由 Weather Panel 气候板(意译：自控光热板)组成。

图 9-6 为该板外形。图 9-7 示出两种剖面：(a)型板，左为室外，自左→右：低发射率涂层，低发射率隔片，浮云状胶体(可调光栅板)，塑料膜或玻璃或玻璃纤维；(b)型板，低发射率涂层，低发射率隔片，可调光栅板，选择性充水贮热层，选择性深色涂层及内饰面。图 9-8 为节点细部。

9.4 Weather Panel Roof

The example roof is constructed by a weather panel (i. e. A panel can self-controlling light and heat).

Fig. 9-6 shows the shape of the panel. Fig. 9-7 shows the panel's two sections: Type (a), left is outside, from left to right: low emissivity coating, low emissivity baffles, cloud gel (adjustable louvers), plastic film or glass or fiberglass; type (b), low emissivity coating, low emissivity baffles, adjustable louvers, selective water thermal storage, selective

9.4 自控光热屋顶
9.4 Weather Panel Roof

dark paint and interior finish. Fig. 9-8 the joint detail.

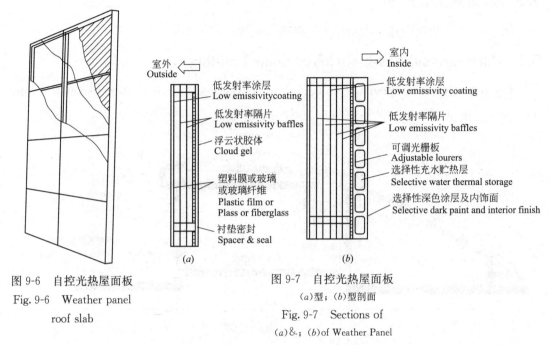

图 9-6 自控光热屋面板
Fig. 9-6 Weather panel roof slab

图 9-7 自控光热屋面板
(a)型；(b)型剖面
Fig. 9-7 Sections of (a)&; (b) of Weather Panel

简言之，自控光热屋面乃是由多层低发射率透明片材构成的多层密闭空气层，并由浮云状胶质随太阳辐射强弱变动透明度调节入室光、热。它是一种轻型、高绝热和有效利用太阳能的屋顶。耐久性可达30～40年。

图9-9为一幢四单元自控光热屋面板太阳能公寓。

In short, the weather panel roof is composited by multiple layers of low emissive transparent films forming a veneered sealed air gaps. It adjusts the in-coming daylight and solar heat by the cloud gel whose transparency is changeable according to the solar radiation strength. This roof is lightweight and high insulation and use solar energy efficiently. Its durability may be between 30～40 years.

Fig. 9-9 shows a solar flat with 4-units using Weather Panel Roof.

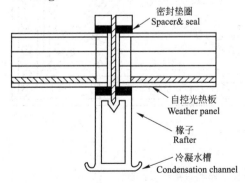

图 9-8 自控光热屋面板节点细部
Fig. 9-8 The joint detail of the Weather Panel

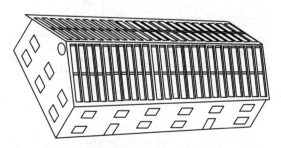

图 9-9 四单元自控光热屋面板太阳能公寓
Fig. 9-9 A solar flat with 4-units using weather panel roof

9 优化组合
9 OPTIMAL COMPOSITIONS

9.5 地面掩土太阳房

图 9-10 为一幢地面掩土太阳房平、剖面。该房已建成多年。

9.5 Aboveground Earth Sheltered Solar Building

Fig. 9-10 shows the plan and section of an aboveground earth sheltered solar building. This building has been built for many years.

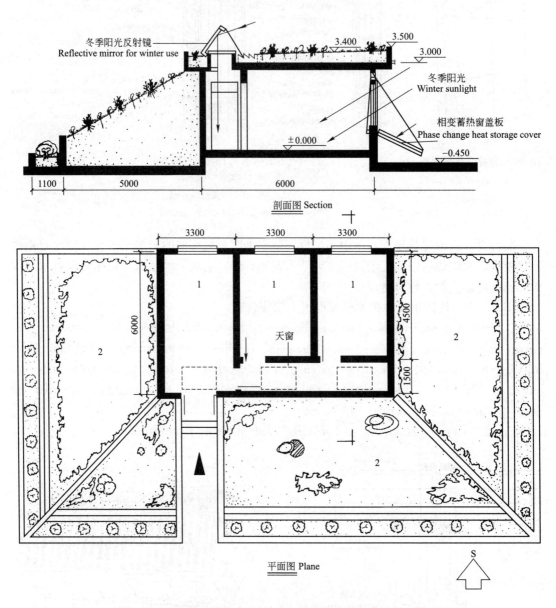

图 9-10 地面掩土太阳房（设计者：成炎、施燕，指导者：夏云）

Fig. 9-10 Aboveground earth sheltered solar building(Designers：Cheng Yan，Shi Yan. Director：Xia Yun)1—办公室 office room 2—覆土种植 Soilcovering & planting

9.6 下沉式太阳能教室

9.6 Sunken Solar Classroom

设计要点:

1) 通过南窗直接得热。

2) 利用反射镜使室内北廊成为太阳能暖廊,光热环境大改善。与一般内廊比,光线改善10~20倍,曾测到照度高达250~300lx,达Ⅱ级采光标准。

3) 东、西、北三面覆土自然绿化,使抗冲击波能力、热阻与贮热效能均得到提高。

4) 屋顶:8cm现浇钢筋混凝土板,2cm水泥砂浆找平层,再是防水涂层。其上为30cm锯末掺黄土(体积比1∶1)自然绿化层。自然绿化效果甚佳。屋顶有的飞来种子长的树已高达4m多(图6-25)。

5) 南窗设有相变贮热窗盖板。相变材料为($Na_2SO_4 \cdot 10H_2O$)(芒硝)。白天吸热、贮热,晚上向室内放热。(十分遗憾,当时(1980年)买到的芒硝竟是假货,致使此项实验未做成)。

Main features of design are:

1) Direct gain through southern windows.

2) Using reflective mirror to make the northern corridor a solar warm corridor with great improvement of light and heat, comparing with normal ones. Lighting level has improved 10~20 times and the highest illuminance has reached 250~300lx, which meets with the second degree of lighting standard.

3) Natural landscaping combined with the cover soil at east, west and north improve the anti-shock wave ability, thermal resistance and heat store efficiency.

4) Roof: consists of 8cm thick in-situ RC slab, 2cm thick cement mortar leveling coat, waterproof coating, and 30cm thick saw-dust mixed loess (volume rate: 1∶1) as natural landscaping layer. It produced a very good landscaping result with some trees grown over 4 metres tall from air-borne seeds (see Fig. 6-25).

5) South facing windows have been designed with covers with phase change heat store material $Na_2SO_4 \cdot 10H_2O$ (Glauber('s) salt). It absorbs and stores heat during the day and heat the room at night. (It was a great pity that this experiment had not been completed because then (1980s) the mirabilite we bought was fake).

9.6 下沉式太阳能教室

图9-11为一下沉式太阳能教室剖面。

设计要点:

1) 地面、墙、屋顶全为现浇钢筋混凝土结构。屋顶防水层上铺30cm蛭石掺黄土(体积比3∶1)种植层,并设置太阳能热水器、太阳能电池板、太阳能灶供日常使用及太阳能知识教学。

2) 自然采光主要由屋顶天窗提供。

3) 为消减马路运输固体传声,采用氯丁橡胶止水带隔振和消声沙坑(图9-11(a)、(b))。

4) 太阳能廊地面及向阳墙裙采用相变贮热构造。

9 优化组合
9 OPTIMAL COMPOSITIONS

9.6 Sunken Solar Classroom

Fig. 9-11 shows a section of the sunken solar classroom.

Main points of design are:

1) Its floors, walls and roof all are RC structure. On top of waterproofing, a layer of mixed vermiculite loess (volume rate: 3:1) for planting is laid. There are solar hot water heaters, solar cell panels and solar cookers for daily use and solar education.

2) Natural lighting is mainly supplied by the skylights.

3) To reduce solid-borne noise from the highway traffic, chloroprene rubber water stop tie and sand pit silencer have been used (Fig. 9-11 (a)、(b)).

4) Solar corridor floor and sun-facing dados use phase change heat store details.

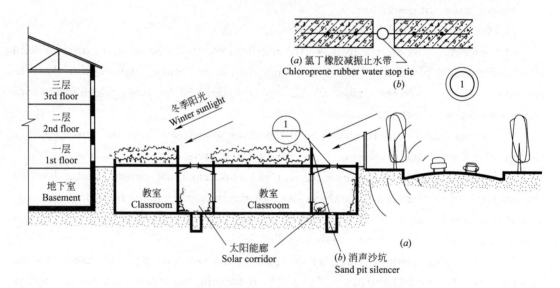

图 9-11 下沉式太阳能教室(设计：黄瀛，指导：夏云)

Fig. 9-11 Sunken solar classroom
(Designer: Huang Ying, Director: Xia Yun)

9.7 下沉式温室自然空调太阳房

图 9-12 为该房剖面。

设计要点：

1) 将传统下沉式窑洞露天天井院改建成透明屋顶的下沉式种植温室。

2) 将地道风自然空调与绿化温室结合起来提高生态效益。

3) 运行原理：

冬季，室外干、冷，鲜空气(FA)经捕风器(W)转向下行，经露点深度(D)(6～8m)由上到下受土壤夏季贮热加热，再经水平风道卵石床(P)稳定所得升温，并加湿。升温增湿后的新鲜空气经水平风道末端由 V_3、V_4 入绿化温室补氧，并二次加温后，经 V_5、V_6 入室(此时，V_1、V_2 应关闭)。经人呼吸含 CO_2 多的热空气由隔污换热器

9.7 下沉式温室自然空调太阳房
9.7 Sunken Greenhouse Solar Building with NAC

(H)内管排出,途中对外管冷鲜进气加热。内管暴露段外表涂黑(符号 SB),形成太阳能烟囱效应增强排气。排气管顶端做成"T"形管帽,可利用各向风力增强抽气(详见 7.3.3)。

夏季,干热或湿热鲜空气进入露点深度(D),被土层吸热降温后入水平风道,在那儿一股鲜空气由开启的 V_1、V_2 入室,另一股鲜空气仍经 V_3、V_4 入温室补充植物光合作用所需 CO_2,而进入的 O_2 则与植物光合作用产生的 O_2 加在一起,使温室含氧增加,并由天窗 V_7、V_8 排出(V_7、V_8 冬季关闭),改善了周围区域的空气质量。

9.7 Sunken Greenhouse Solar Building with NAC

Fig. 9-12 is a section of the building.

Main points of design:

1) To make the conventional sunken cave's open court a sunken plantable greenhouse with a transparent roof.

2) The tunnel ventilation from NAC combining plantable greenhouse can increase eco-efficiency.

3) Operation principles:

In winter, the dry, cold and fresh outdoor air (FA) is drawn down to the Dew Point Depth (D) (6~8m) by a wind-catcher (W). The air will be warmed up by the heat stored in the soil during summer as it travels from top to the bottom of the tunnel. Then the warmed air passes through pebble bed in horizontal tunnel (P) to stabilize its temperature and carry more moisture. The warm, moist fresh air then goes through the horizontal tunnels' end vents V_3 and V_4 to enter into the planting greenhouse to receives oxygen supplement and be heat up once again. It is finally time for the air to go into the rooms via vents V_5 and V_6 (while the V_1、V_2 should be closed). Through human breath the warm air containing more CO_2 is to be discharged from the inner tube of the pollution isolation heat exchanger (H). During the process, the out going air will heat in coming cold fresh air. The external face of the exposed part of the inner tube is painted black (Mark: SB). It utilizes solar chimney effect to strengthen air discharge. The "T" shaped end cap of the inner tube can use wind from any direction to strengthen air discharge too (detail in 7.3.3).

In summer, hot-dry or hot-wet fresh air goes into the Dew Point Depth (D). Its heat will be absorbed and its temperature lowered. It then goes into the horizontal tunnel at where a part of the fresh air goes into the rooms via opened vents V_1 and V_2. Another part of fresh air goes into the greenhouse via vents V_3 and V_4 to supplement CO_2 for the plants' photosynthesis. The in coming O_2 together with the O_2 generated by the photosynthesis will make the greenhouse oxygen-rich and be discharged via the skylights V_7 and V_8 (in winter, V_7 and V_8 should be closed) to improve surrounding air quality.

9 优化组合
9 OPTIMAL COMPOSITIONS

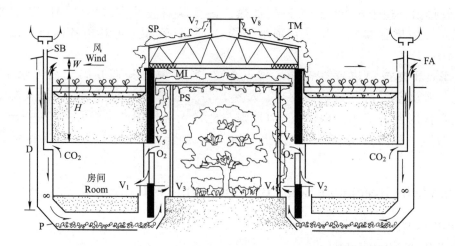

图 9-12 下沉式温室自然空调太阳房，"∞"太阳能电池风扇
（设计：夏葵 陈晓育，绿化设计：任云英）

Fig. 9-12 Sunken greenhouse solar building with NAC, "∞" solar cell fan
(Designers: Xia Kui, Chen Xiaoyu; Greening designer: Ren Yunying)

图 9-13 是一幢两层楼带下沉式毗连温室的自然空调太阳房剖面。本建筑运行原理与图 9-13 建筑基本相同，其不同处：图 9-12 水平风道末端有开孔，当风道内有凝结水时可通过开孔渗入土层，成为地下水。当风道内需对干空气加湿时，通过开孔可取得地下毛细水作为湿源。显然图 9-12 建筑适用于地下水最高水位低于水平风道的地区，如陕、甘、宁、青、新、豫、藏、内蒙古大部地区以及占国土面积 2/3 以上的各种山坡地带（21 世纪发展山地建筑决不可忽视）。这些地区地下水位常在离地面 10～20m，甚至 100～200m 之多。图 9-13 中水平风道无开孔，适用于各种地区含地下水位高于水平风道地区。此处，地下房间、毗连温室、风道等均应作防水处理。湿热地区，水平风道可能有冷凝水，可由与毗连温室相通的开口流入温室收集井。干热地区可在自然空调进气口设喷雾龙头对进气加湿（图 9-16），或用毗连温室集水井作湿源。

Fig. 9-13 shows the design section of a two-story solar building with an attached sunken greenhouse and NAC. This building operates basically the same as the building showing in Fig. 9-13. The difference between the two is that: in Fig. 9-12 the end of horizontal tunnel has opening to allow possible condensation in the tunnel penetrates through the opening to become ground water. When need to add moisture to the in coming dry air in the tunnel, capillary water can be drawn through the openings. Obviously, the horizontal tunnel with opening is suitable to where the highest ground water table is lower than the level of the horizontal tunnel. These areas are generally located in most part of Shaanxi, Gansu, Ningxia, Qinghai, Xinjiang, Henan, Tibet, Inner Mongol and various hillsides areas. The overall areas are more than 2/3 area of the country. In the 21st century, we should pay more attention to develop hillside buildings, where the highest ground water table is usually 10～20m below ground and some even 100～200m below. In Fig. 9-13 no opening is at the horizontal tunnel. It's suitable to where the ground water ta-

9.8 零—零建筑(双零建筑)
9.8 Zero—zero Building (Double Zero Building)

ble is higher than the level of the horizontal tunnel. The underground rooms, attached greenhouse, tunnel, and etc. in these areas should be waterproofed. In hot-wet region, possible condensation in the horizontal tunnel may be flown into the collecting well in the attached greenhouse through the opening between them. In hot-dry region, a sprayer at the air inlet of the NAC should be installed to add moisture to the in coming air (Fig. 9-16) or use the collecting well in the attached greenhouse as the source of moisture.

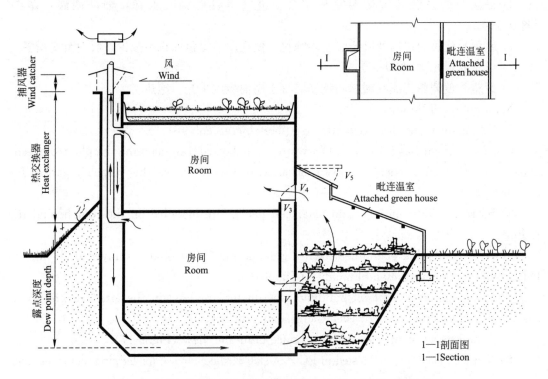

图 9-13 下沉式毗连温室自然空调太阳房(设计：夏葵，陈晓育)
Fig. 9-13 An attached sunken greenhouse solar building with NAC
(Designers：Xia Kui，Chen Xiaoyu)

9.8 零—零建筑(双零建筑)

本建筑不用传统污染性能源，特称"零能"建筑；又不损失绿地面积，特称"零地"建筑，合称"零—零建筑"或"双零建筑"。图 9-14 为设计模型照。图 9-15 为部分平、立、剖面。地址为兰州。

9.8 Zero—zero Building (Double Zero Building)

This type of building uses no conventional energy. So it can be called "zero energy building". Besides, it losses no landscaped area hence be called "zero land building". Combining both characters, we call it "Zero-zero building" or "Double zero building". Fig. 9-14 shows the photograph of a design model. Fig. 9-15 shows parts of its plans, facades and

9 OPTIMAL COMPOSITIONS

sections. Site：Lanzhou.

设计要点：

1) 采用毗连种植温室太阳房

我们实验证实，毗连种植温室冬季晴天白天可节能50%甚至更多，同时它也是每户的全天候毗连花园。

2) 屋顶设太阳热水器、太阳电池板于防护栏杆后。

3) 自然空调系统与种植温室相结合。进气口设可调喷水管，辅助调温，调湿（图 9-16）。

4) 屋顶及建筑周边以及毗连温室均绿化，使建房后绿色面积不减少，反而有所增多。

5) 设实验引光廊（图 9-17）。

6) 废罐头玻璃瓶浇灌在楼梯间钢筋混凝土屋面板内采光、绝热。

Main points of design：

1) Using solar building with attached plantable greenhouse

Our experiment has proved that in winter fine days, during daytime the planted greenhouse can save 50% of energy or even more. It is also an all-weather attached garden for each family.

2) Solar hot water heaters and solar cell panels are installed on the roof behind the railing.

3) NAC combining plantable attached greenhouse, a controllable sprayer at air inlet to help regulate temperature and humidity (Fig. 9-16).

4) Landscaping on roof, on buildings' perimeter area and in the attached greenhouses, make no loss green area or even increase after it's built.

5) Install experimental shaft to introduce light and view (Fig. 9-17).

6) Recycled glass jars are casted in RC roof slab of the staircase for lighting and insulation.

图 9-14 双零建筑模型照片

Fig. 9-14 A photograph of the doublezero building model

9.8 零—零建筑(双零建筑)
9.8 Zero—zero Building (Double Zero Building)

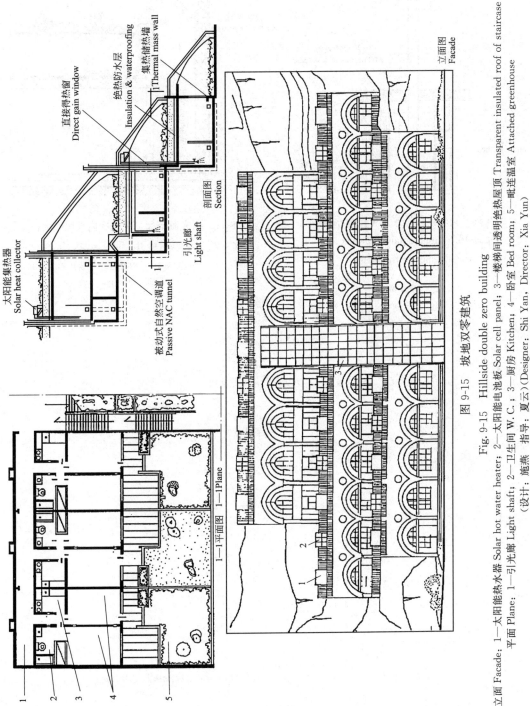

图 9-15 坡地双零建筑
Fig. 9-15 Hillside double zero building
立面 Facade: 1—太阳能热水器 Solar hot water heater; 2—太阳能电池板 Solar cell panel; 3—楼梯间透明绝热屋顶 Transparent insulated roof of staircase
平面 Plane: 1—引光廊 Light shaft; 2—卫生间 W.C.; 3—厨房 Kitchen; 4—卧室 Bed room; 5—毗连温室 Attached greenhouse
(设计：施燕　指导：夏云)(Designer: Shi Yan, Director: Xia Yun)

9 优化组合
9 OPTIMAL COMPOSITIONS

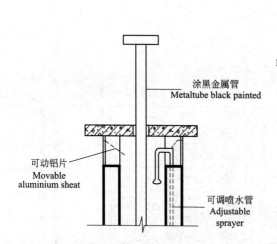

图 9-16 自然空调进气口可调喷水器

Fig. 9-16 An adjustable sprayer at the air inlet of the NAC

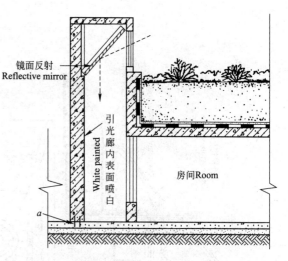

图 9-17 引光廊剖面，引光廊内表面喷大白

Fig. 9-17 Section of the light shaft, its inner surface are painted white

a—插筋 Dowel bar

10 生态可持续建筑的经济效益
10 ECONOMIC BENEFITS OF ESA

10.1 回收与净得益

生态可持续建筑第一次投资比传统建筑会多些。但由于有节能等回收效益，据已有经验，一般5~8年即可完全回收，以后就进入净得益年了。图10-1横坐标表示回收期、净得益期与房屋使用年限，纵坐标表示投资。设传统建筑投资如A线，生态可持续建筑投资多，如B线所示。由于节能等效益，该建筑按生态可持续建筑建造则可得到一回收累积增长率C线。C与B交于N，ON投影ON′即回收期年限，以后即净得益累积增长期。若房屋使用50~100年，该生态可持续建筑至少就有40~90年的净得益年！与此同时，传统建筑却处在耗能等支出累积增长期。此外，生态可持续建筑的生态效益更是很难用金钱衡量。

10.1 Benefit and Net Gain

The initial investment of ESA is more than that of conventional building, but ESA has long term benefits from energy saving and other measures. Based on the experience, in 5 to 8 years the first investment can be all paid back. From then on, it will go into net gain years. In Fig. 10-1, the horizontal line indicates the pay back period, net gain period and life span of the building. The vertical line shows the amount of investment. Assume line A is the investment of a conventional building and line B is the investment of an ESA which is more than Line A. Because of the benefits of energy saving and other ESA measures, the

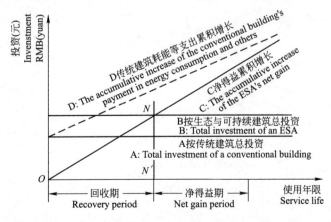

图10-1 传统建筑与生态可持续建筑投资效益比较

Fig. 10-1 A comparison of the investment benefits between conventional building and ESA

10 生态可持续建筑的经济效益
10 ECONOMIC BENEFITS OF ESA

ESA can get a accumulative net pay back rate as Line C. Line C crosses Line B at point N. Line ON's projection ON′ on the horizontal line is the pay back time. From then on, it's going into the net gain period. If the building life span lasts between 50~100 years, the ESA will at least have 40~90 years of net gain period! While the conventional building will still be in the accumulative pay back period. Besides, the ESA's eco-benefits are too hard to be calculated in money terms.

主要参考文献
MAIN CONSULT DOCUMENTS

[1] 吴良镛主笔. 北京宪章. 北京：国际建协(UIA)第20届大会，1999(会议文献)
[2] 李博. 普通生态学. 呼和浩特：内蒙古大学出版社，1990
[3] 张鹤飞. 太阳能热利用原理与计算机模拟. 西安：西北工业大学出版社，1990
[4] C. B. WILSON. BUILDING AND ENVIRONMENT (The International Journal of Building Science and Applications). PERGAMON PRESS, 1994. 1
[5] Lawrence L. Kazmerski. RENEWABLE&SUSTAINABLE ENERGY REVIEWS. PERGAMON PRE—SS, March. 1999
[6] S. D. PROBERT&P. WALSH. APPLIED ENERGY. ELSVIER. 1998
[7] 熊明等. 城市设计学. 北京：中国建筑工业出版社，1999
[8] 周若祁等. 绿色建筑. 北京：中国计划出版社，1999
[9] 王朝晖. 中国可持续建筑理论框架与适用技术的探讨：[博士学位论文]. 北京：清华大学建筑学院，1999(学位论文)
[10] 宋晔皓. 结合自然整体设计注重生态的建筑设计研究：[博士学位论文]. 北京：清华大学建筑学院，1998(学位论文)
[11] (日)八代克彦. 中国黄土高原の下沉式窑洞住居にぉけゐ中庭空间の配置构成に间るず研究：[博士论文]. 日本東京：東京工业大学，1993(学位论文)
[12] 代彦军. 干燥剂除湿与蒸发冷却复合传热传质及应用研究：[博士论文]. 西安：西北工业大学，1999(学位论文)
[13] 蔡宏道等. 中国医学百科全书. 环境卫生学. 上海：上海科技出版社，1982
[14] 李莉萍. 大理、丽江民族建筑与人聚环境持续发展的研究. 昆明：云南工业大学建筑系，1996
[15] 吴红. 城市地下建筑设计研究：[硕士学位论文]. 南京：东南大学建筑系，1998(学位论文)
[16] 王宏经. 何健. 中国当代建筑论坛. 济南：山东大学出版社，1997
[17] 夏葵. 夏云. 21世纪建筑技术科学何往. 上海：全国建筑技术学科第八次代表大会暨学术讨论会，2000(会议文献)
[18] 柳孝图. 建筑物理. 北京：中国建筑工业出版社，1991
[19] 邱凌等. 沼气发酵与综合利用. 西安：天则出版社，1990
[20] 夏云. 王进. 夏葵译. 建筑科学基础(澳)Jack Greenland 著. 西安：陕西科学技术出版社，1996
[21] 夏云译. 掩土建筑—历史、建筑与城镇设计(美)吉·戈兰尼著. 北京：中国建筑工业出版社，1987
[22] 夏云. 夏葵. 节能节地建筑基础. 西安：陕西科学技术出版社，1994
[23] 黄瀛. 太阳能廊在浅层地下空间中的应用. 北京：建筑知识，1991，(3)：22～23
[24] 夏云. 夏葵. 生态建筑与建筑的持续发展. 北京：建筑学报，1995. 6：6～9
[25] 夏云. 王进. 21世纪中国节能节地建筑展望. 北京：建筑学报，1991. 10：47～51
[26] Xia Yun, Wang Jin. NEW POSITION OF GREENHOUSE. Beijing：Proc. International Greenhouse Technology Symposium, 1991, 20～24
[27] Xia Yun, Wang Jin, Xia Kui. SOLAR BUILDING WITH NATURAL AIR CONDITIONING. Yugoslavia：Proc. The 11th International Heating, Cooling and Air Conditioning Symposium, 1991, 228～232

主要参考文献
MAIN CONSULT DOCUMENTS

[28] Zhang Yiren, Xia Yun. COMPREHENSIVE TREATMENTS: The Optimal Selection of Energy and Land-saving Buildings in China. Beijing: Proc. International Conference on New and Renewable Energy, 1990, 408~411

[29] Xia Yun, Mao Qinhong, Zhang Yiren. MULTI-USE OF SOLAR ENERGY IN A BUILDING. Beijing: Proc. International Conference on New and Renewable Energy, 1990, 419~424

[30] Helen Mulligan, Xia Yun, Zhang Yiren. THE CHINESE CAVE DWELLING: A Solar store. Lanzhou, China: Proc. United Nations Symposium on Solar Energy Applications in Developing Countries, 1989, 199~202

[31] Xia Yun, Helen Mulligan, Zhang Yiren. SOLAR BUILDING CHALLENGES CONVENTIONAL BUILDINGS. Lanzhou, China: Proc. United Nations Symposium on Solar Energy Applications in Developing Countries, 1989, 203~207

[32] Xia Yun, Zhang Yiren. COMPREHENSIVE AND LOW ENERGY ARCHITECTURE IN NORTHWESTERN CHINA. Kobe, Japan: Proc. ISES Solar World Congress, 1989, 1124~1128

[33] Xia Yun, Zhang Yiren. GO UNDERGROUND TO GET BUILDING SPACE. Shanghai, China: Proc. The Third International Conference on Underground Space and Earth Sheltered Buildings, 1988, 7~10

[34] Shi Yan. DEVELOPING EARTH SHELTERED BUILDINGS TO IMPROVE LANZHOU CITY'S DEVELOPMENT. Shanghai, China: Proc. The Third International Conference on Underground Space and Earth Sheltered Buildings, 1988, 43~44

[35] Xia Yun, Zhen Yan, Shi Yan. EXPERIENCES IN USING EARTH SHELTERED BUILDINGS AND UNDERGROUND SPACES IN SHAANXI CHINA. Minneapolis, USA: Proc. Second International Earth Sheltered Buildings Conference, 1986, 41~44

[36] Xia Yun et al. AN ENERGY SAVING AND LAND SAVING EXPERIMENT ON THE LOESS CAVE IN CHINA. Melbourne, Australia: Proc. Owner-Building and Earth Architecture, 1984, 242~251

[37] Xia Yun, Zhang Yiren. OPTIMUM SYSTEM OF SOLAR ARCHITECTURE. Budapest, Hungary: Proc. ISES Solar World Congress, 1993, 143~148

[38] Xia Yun, Zhang Yiren. ENERGY AND LAND—SAVING BUILDINGS IN CHINA. Moscow: Lecture at Research Institute of Russia Village Electrization, 1994

[39] Xia Yun. CHINESE UNDERGROUND SPACE AND NATURAL AIR CONDITIOING. Lecture at Ecole d'Architecture de Bordeaux and Ecole d'Architecture de Nantes, France, 1995

[40] 夏云. 生态与可持续建筑. 对清华大学建筑学院, 天津大学建筑学院等11所院校讲学手稿, 1998~2000

[41] 夏云. 生态与可持续建筑. 对西安建筑科技大学建筑学院该课程讲稿, 1995~2009

[42] 夏云. 夏葵. 施燕. 生态与可持续建筑. 北京: 中国建筑工业出版社, 2001

[43] 夏云. 生态与可持续发展建筑. 北京: 建筑学报, 2002. 7: 8~10

[44] 夏云. Renewed Chinese Cave Dwellings-Sustainable Earth-sheltered Houses 更新中国窑洞住宅—可持续发展掩土房. 台北: 建筑杂志, 068 APR2003: 134~141

[45] 靳亦冰. 中国西部都市可持续发展必由之路. 西安: 亚洲都市环境的可持续发展国际学术会议, 2005. 11

[46] 陈洋. 论中国高校生态可持续校园模式: [博士学位论文]. 西安·西安建筑科技大学, 2004

[47] 陈洋. 关于黄土高原住区的集雨节水. 《华中建筑》, 2001

[48] 陈晓育. 夏云. 中国乡土建筑绿色之途. 2008绿色建筑与建筑新技术国际会议论文集, 82~87, 北京: 中国建筑工业出版社

[49] 夏云. 陈晓育. 中国现代 城市建设及其科技、文化传承. 西安: 建筑与文化2008国际学术讨论会,

主要参考文献
MAIN CONSULT DOCUMENTS

2008. 11
- [50] 夏云. 夏葵. 建筑科学基础 FOUNDATIONS OF ARCHITECTURAL SCIENCE(汉-英双语). 北京：建材工业出版社，2005
- [51] 夏云. 靳亦冰. 再论生态可持续建筑. 北京：建筑学报，2009 学术论文专刊 01
- [52] 夏云. 重大天灾人祸下对建筑的反思与对策. 2009. 4 专题报告，西安建筑科技大学研究生院组织主持
- [53] 住房和城乡建设部科技发展促进中心等. 绿色建筑的人文理念. 北京：中国建筑工业出版社，2010
- [54] 动感·生态城市与绿色建筑编辑部. 生态城市与绿色建筑(ECO-CITY AND GREEN BUILDING)创刊号. 北京：中国出版期刊中心，2010
- [55] 中国建材科研院. 绿色建材与建材绿色化. 北京：化学工业出版社，2003
- [56] 刘先觉等. 生态建筑学. 北京：中国建筑工业出版社，2009
- [57] 俞孔坚. 回到土地. 北京：生活·读书·新知 三联书店，2009